SOLID STATE DIFFUSION

J. P. STARK

Department of Mechanical Engineering
The University of Texas at Austin

ROBERT E. KRIEGER PUBLISHING COMPANY
MALABAR, FLORIDA
1983

Original Edition   1976
Reprint Edition   1983

Printed and Published by
ROBERT E. KRIEGER PUBLISHING COMPANY, INC.
KRIEGER DRIVE
MALABAR, FLORIDA 32950

Library of Congress Cataloging in Publication Data

Stark, John Paul, 1938-
     Solid state diffusion.

        Reprint of the ed. published by J. Wiley, New York.
     Includes bibliographical references and index.
        1. Dilute alloys. 2. Diffusion. I. Title.
[TN690.S75   1983]   669'.94   80-11750
ISBN 0-89874-145-9   AACR1

SOLID STATE DIFFUSION

To Susan Anne

PREFACE

This graduate level book presents the theory of diffusion in solids in as logical a manner as is known to this author. I have included the composite results of many approaches to the problems of diffusion. Much of the understanding is based on varied topics that bear on the subject. Where possible, I have stressed the kinetic theory with examples of dilute alloys. Often, however, I have resorted to the phenomenological developments, because of the complexity associated with the straightforward random walk and kinetic theory approaches.

With the exception of the first chapter, the mathematical background necessary includes only partial differentiation and some matrix algebra. The latter subject is often missed by beginning materials science and metallurgy graduate students; however, the depth herein is not overwhelming since the inverse matrix is the only real complexity used. Chapter 1 includes a fairly complete discussion of the operations surrounding its use and some complex variables, neither of which should present the student with too much difficulty.

Chapters 2, 3, and Appendix A contain some statistical mechanics. The level is comparable to that of *Introduction to Statistical Thermodynamics* by T. Hill, which is highly recommended. The maximum term method is used extensively in the three places where statistical mechanics is necessary. The principle behind the maximum term method is, of course, the minimum free energy of an equilibrium state.

Some readers may feel that the extensive kinetic development in Chapter 2 is presented too early in the text. In my view, it is the most logical way of introducing the phenomenological theory; that is, by showing that the diffusion flux is proportional to the chemical potential gradient, rather than by assuming it to be the case based on some vague thermodynamic presumptions.

The phenomenological theory is of great value in understanding the experiments in concentrated systems. Unfortunately, some authors have not been sufficiently careful in their definitions. Consequently, self-diffusion is carefully defined in Chapter 4, and subsequently related to off-diagonal phenomenological coefficients and vacancy wind terms with specific examples for dilute alloys. The same set of phenomenological coefficients arise when external fields influence mass transfer. This gives the opportunity to compare results from both sets of experiments or theories, as is done in that chapter.

There are problems included for all chapters. The easier problems are found in the earlier chapters; difficult problems and those that require a computer are marked with a dagger. The problems are included to supplement the text in depth and scope.

J. P. STARK

Austin, Texas
February 1976

CONTENTS

SOLID STATE DIFFUSION

CHAPTER 1

INTRODUCTION

FLUX EQUATIONS FOR A CONTINUUM

The motion of atoms in crystals follows various laws that can be written for a macroscopic continuum to follow the dissolution process. The connection between the continuum behavior on which experiments are based and the atomic behavior that can explain the continuum laws is the major purpose of research in diffusion. This is not to imply that the continuum laws are unimportant; they are the basis for nearly all the studies of phase transformations, oxidation rates, and other phenomena related to mass transport in the solid state. However, an understanding of the parameters that are present in the continuum laws and that control the rates with which these processes occur provides part of the basis for the innovative development of new materials behavior.

The manipulation of the continuum laws is usually an exercise in applied mathematics. It may be assumed that such a topic is irrelevant to the presentation of diffusion in solids. This may be true for a large number of cases; however, in research one is attempting to interpret the continuum behavior in terms of the kinetics of atom motion, and such an interpretation can be prejudiced by an incorrect application of the continuum laws. This is particularly true if, for example, two different interpretations would lead to the same results within experimental error. As a consequence, in this book some attention is paid to the manipulation of these laws in the form of differential equations.

First, however, the continuum equations are introduced. The continuum results will ultimately be compared to the kinetic behavior of

1

individual atoms. These atoms move relative to the crystallographic planes on which the atoms reside, and consequently, the continuum behavior should be written relative to a coordinate system that moves at the same velocity as these atom planes. The flux of matter of a given component in a homogeneous and multicomponent system is a vector quantity reflecting the mass of that component crossing a unit area per unit time. The unit area is normal to the flux vector. An equation for the flux of component K, J_K is

$$J_K = -D_K^* \nabla C_K + C_K V_{FK} \qquad\qquad 1.1$$

where D_K^* is the self-diffusion coefficient of a tracer of component K, C_K is the concentration of component K, and V_{FK} is the drift velocity of component K due to a force F. The diffusion coefficient is generally a tensor of rank 2 with typical units of cm^2/sec. The concentration is in some appropriate unit of mass per unit volume, such as $moles/cm^3$. The product of the diffusion coefficient tensor and the concentration gradient is a tensor contraction.

When the velocity due to an applied force V_{FK} in Eq. 1.1 is zero, this result is known as Fick's first law. The occurrence of the velocity term results from the incidence of atoms experiei ing a net drift due to some applied field. Some of the fields that are experimentally known to cause mass transfer are the electric field, gravitational field, temperature gradient, chemical potential gradient, and so forth.

If the lattice planes are moving at some velocity V_L relative to the fixed laboratory frame of reference, then in this fixed coordinate system the flux, J_K^L, is related to the flux in Eq. 1.1 by the relationship

$$J_K^L = J_K + C_K V_L = -D_K^* \nabla C_K + C_K (V_{FK} + V_L) \qquad 1.2$$

Such a situation is often the case because the applied fields causing the motion of component K will also cause the drift of all other components in the system, and if each component moves relative to the lattice at some different rate, the lattice itself may move.

The net rate of change of the mass of component K within a given volume fixed relative to the laboratory coordinates is

$$\frac{d}{dt} \int^V C_K \, dV = \int^V \frac{\partial C_K}{\partial t} \, dV \qquad\qquad 1.3$$

The mass of component K inside the volume element V can change in two ways. First, by Eq. 1.2 one can have flow through the surface of the volume element. Second, one can produce an annihilate component K through a chemical reaction, although this is less common in solids. Generally, there may be r chemical reactions of which each contributes to the production of component K. Let the reaction rate per unit volume be $\nu_{Kj}R_j$ for the jth chemical reaction. The rate of production of K per unit volume from the various chemical reactions is

$$\sum_{j=1}^{r} \nu_{Kj}R_j$$

where ν_{Kj} is positive for production and negative for annihilation of component K. The coefficients ν_{Kj} divided by the molecular mass M_K is proportional to the stoichiometric coefficient with which component K is found in the jth chemical reaction, and R_j is the rate that reaction j proceeds in units of mass per volume x time. The surface of the volume element, Ω, yields a vector $d\Omega$ whose normal direction points outward from the volume. The reaction rates and the flux contribute to the rate of change in the concentration of component K through the relation

$$\frac{d}{dt}\int^{V} C_K\,dV = -\int^{\Omega} J_K{}^L \cdot d\Omega + \sum_{j=1}^{r}\int^{V}\nu_{Kj}R_j\,dV \qquad 1.4$$

Gauss' theorem may be applied to the surface integral; this leaves Eq. 1.4 as

$$\frac{d}{dt}\int^{V} C_K\,dV = \int^{V}\left(-\nabla \cdot J_K{}^L + \sum_{j=1}^{r}\nu_{Kj}R_j\right) dV$$

$$= \int^{V}\frac{\partial C_K}{\partial t}\,dV$$

However, the volume element V is completely arbitrary, and this implies that

$$\frac{\partial C_K}{\partial t} = -\nabla \cdot J_K{}^L + \sum_{j=1}^{r}\nu_{Kj}R_j \qquad 1.5$$

Equation 1.5 is a general statement of the law of conservation of mass.

One might argue that it is extraneous to consider chemical reactions occurring within a homogeneous crystal. A simple example can show the importance of such terms in the conservation equation 1.5. Consider, for example, the motion of hydrogen through a steel bar that is about to be embrittled. The hydrogen probably goes into solution in a monatomic form as it resides on interstitial sites within the crystal. If a hydrogen ion meets an oxygen ion and they are close enough together, they may share some of the surrounding electrons forming a pair somewhat of the form $H:O$. This pair will move through the crystal at a different rate than does an isolated hydrogen ion: hence there may be two important diffusion coefficients to consider, that of the pair and that of the isolated hydrogen ion. It may be convenient in some cases to consider the pair as being less mobile, and such trapping mechanisms may decrease the rate of embrittlement of the steel sample by hydrogen.

A specific example of a problem, in which it was convenient to use chemical reactions in solids, is discussed in Chapter 7. At this point, however, assume that $R_j = 0$ for all j; there are no chemical reactions. Furthermore, in most research problems, Eq. 1.1, 1.2, and 1.5 are used in a one-dimensional form rather than three dimensions as indicated. This is true because the research experimenter attempts to determine a kinetic interpretation of the tracer diffusivity, D_K^*, and the velocity from an applied force, V_{FK}, for each component. This information, as well as a solution to the differential equation 1.5, is most reasonably accomplished with the concentration gradients and applied force collinear in some crystallographic direction. If only the diagonal elements of the diffusion coefficient, written as a matrix, are nonzero (as shown for some systems in this chapter), and assuming the applied field and concentration gradients exist only along the x axis Eq. 1.1 becomes

$$J_K = -D_K^* \frac{\partial C_K}{\partial x} + C_K V_{FK} \qquad\qquad 1.6$$

Similarly Eq. 1.5 becomes

$$\frac{\partial C_K}{\partial t} = \frac{-\partial J_K{}^L}{\partial x} \qquad\qquad 1.7$$

$$= D_K^* \frac{\partial^2 C_K}{\partial x^2} + \frac{\partial D_K^*}{\partial x} \cdot \frac{\partial C_K}{\partial x} - \frac{\partial C_K}{\partial x}(V_{FK} + V_L)$$

$$-C_K \frac{\partial(V_{FK} + V_L)}{\partial x} \qquad\qquad 1.8$$

It will become apparent that if one may write the velocity for an applied force as

$$V_{FK} = V_K' - \frac{\partial D_K^*}{\partial x}$$

then it is possible to write Eq. 1.6 as

$$J_K = -\frac{\partial D_K^* C_K}{\partial x} + C_K V_K' \qquad\qquad 1.9$$

If a detailed analysis shows that the tracer diffusion coefficient is position dependent, as would be the case in a temperature gradient, then that analysis would also show that the tracer drift velocity V_K' includes a term $\partial D_K^*/\partial x$. In that instance, Eq. 1.9 must be used in conjunction with Eq.

1.7 because the diffusion coefficient gradient terms do not contribute to the flux; they cancel it. The velocity term V_{FK} in Eqs. 1.1, 1.6, and 1.8 does not contain any contribution from the diffusion coefficient gradient. The contribution of the diffusion coefficient gradient to the flux is discussed in Chapter 2 where a kinetic analysis derives Eq. 1.9 from which Eq. 1.6 may be found by cancelling the appropriate terms.

DIFFUSION IN AN INFINITE CRYSTAL

For an infinitely dilute solute in a binary metal alloy moving through the crystal with an applied electric field, the drift velocity per unit force is nonzero. The net drift velocity is written as $V_{FK} + V_L = Q_K E D_K^*/K_B T$, where Q_K is the effective ionic charge on the solute ion, E is the electric field intensity, K_B is Boltzmann's constant, and T is the absolute temperature. If one assumes that the diffusion coefficient is constant, which will be true if the temperature is uniform, and that E is constant, then Eq. 1.8 becomes

$$\frac{\partial C_K}{\partial t} = D_K^* \left[\frac{\partial^2 C_K}{\partial x^2} - \alpha \frac{\partial C_K}{\partial x} \right] \qquad 1.10$$

where $\alpha = Q_K E/K_B T$. Equation 1.10 is a linear differential equation with constant coefficients. One may solve Eq. 1.10 for the concentration of solute C_K dependent on x and t, provided that appropriate boundary and initial conditions are applied. The most appropriate method of solution for such problems is through the use of integral transforms. We will assume that a thin layer of solute, Q_0 being the mass per unit area, is deposited on the end of the semi-infinite sample of uniform cross sectional area at $t = 0$. At this time the solute is not found elsewhere in the sample. Therefore, the appropriate conditions are

$$C_K(x, t = 0) = 0,$$

$$\int_0^\infty C_K(x, t)\, dx = Q_0,$$

and $\lim_{x \to \infty} C_K(x, t)$ is finite. The method of solution is as follows. Assume that

$$\bar{C}(x, s) = \int_0^\infty C_K(x, t) e^{-st}\, dt \qquad 1.11$$

Equation 1.11 transforms the partial differential equation into an ordinary differential equation. With Eq. 1.11 and integration by parts, it is easy to show that

$$s\bar{C}(x, s) - C_K(x, t = 0) = \int_0^\infty e^{-st} \frac{\partial C_K}{\partial t} dt \qquad 1.12$$

and it is obvious that

$$\frac{\partial^n \bar{C}}{\partial x^n} = \int_0^\infty e^{-st} \frac{\partial^n C_K}{\partial x^n} dt \qquad n = 1, 2, \cdots \qquad 1.13$$

Substituting Eqs. 1.11–1.13 into Eq. 1.10 and using the initial condition that $C_K(x, t = 0) = 0$, Eq. 1.10 becomes

$$\frac{d^2 \bar{C}}{dx^2} - \alpha \frac{d\bar{C}}{dx} - \frac{s}{D_K^*} \bar{C} = 0 \qquad 1.14$$

The Laplace transform, Eq. 1.11, has reduced the partial differential equation, 1.10, into an ordinary differential equation, 1.14. Equation 1.14 is solved by assuming a solution of the form $\bar{C} = A e^{\gamma x}$. With this substitution one finds, assuming $\alpha \gtrless 0$, that

$$\gamma^2 - \alpha\gamma - \frac{s}{D_K^*} = 0 \qquad 1.15$$

However

$$\gamma = \frac{\alpha}{2} \pm \frac{1}{2} \sqrt{\left(\alpha^2 + \frac{4s}{D_K^*}\right)}$$

so that

$$\bar{C} = A_1 \exp\left\{\left[\frac{\alpha}{2} + \frac{1}{2}\sqrt{\left(\alpha^2 + \frac{4s}{D_K^*}\right)}\right]x\right\}$$

$$+ A_2 \exp\left\{\left[\frac{\alpha}{2} - \frac{1}{2}\sqrt{\left(\alpha^2 + \frac{4s}{D_K^*}\right)}\right]x\right\} \qquad 1.16$$

When $\lim_{x\to\infty} C_K(x, t)$ is finite, one infers that $\lim_{x\to\infty} \bar{C}(x, s)$ is finite, from which one concludes that $A_1 = 0$. The other boundary condition,

$$\int_0^\infty C_K(x, t)\, dx = Q_0$$

must be transformed to $\bar{C}(x, s)$ as follows:

$$\int_0^\infty e^{-st} \int_0^\infty C_K(x, t)\, dx\, dt = \int_0^\infty \bar{C}(x, s)\, dx$$

$$= \int_0^\infty e^{-st} Q_0\, dt = \frac{Q_0}{s}$$

$$= \int_0^\infty A_2 \exp -\left\{ \left[\frac{-\alpha}{2} + \frac{1}{2}\sqrt{\left(\alpha^2 + \frac{4s}{D_K^*}\right)} \right] x \right\} dx$$

$$= \frac{A_2}{\dfrac{-\alpha}{2} + \dfrac{1}{2}\sqrt{\left(\alpha^2 + \dfrac{4s}{D_K^*}\right)}} \qquad 1.17$$

Hence

$$A_2 = \frac{-Q_0}{s}\left[\frac{\alpha}{2} - \frac{1}{2}\sqrt{\left(\alpha^2 + \frac{4s}{D_K^*}\right)} \right]$$

According to LaPlace transform theory, the concentration $C_K(x, t)$ is found from the following integral in the complex plane,

$$C_K = \frac{Q_0}{4\pi i}\int_{\sigma - i^\infty}^{\sigma + i^\infty} e^{st} \frac{\left(\alpha - \sqrt{\alpha^2 + \dfrac{4s}{D_K^*}}\right)}{s} \exp\left[\frac{\alpha}{2} - \frac{1}{2}\sqrt{\left(\alpha^2 + \frac{4s}{D_K^*}\right)} \right] x\, dx$$

$$1.18$$

In Eq. 1.18, σ is to the right of all singularities in the function inside the integral. An analysis of the type of singularities present is aided by the substitution $4Z/D_i^* = \alpha^2 + 4s/D_K^*$ and $dZ = ds$. Then

$$C_K = \frac{-Q_0}{4\pi i}\int_{\sigma - i^\infty}^{\sigma + i^\infty} \frac{\exp\left[\alpha x/2 - \alpha^2 D_K^* t/4\right](\alpha - 2\sqrt{(Z/D_K^*)})\exp\left(Zt - x\sqrt{Z/D_K^*}\right) dZ}{(Z - D_K^* \alpha^2/4)}$$

There is a simple pole at $Z - D_K^* \alpha^2/4 = 0$, a branch point at $x\sqrt{Z/D_K^*} = 0$ and at $(\alpha - 2\sqrt{Z/D_K^*}) = 0$. One can get rid of the latter branch point by the substitution $Z = (iV)^2$, and $dZ = -2V\, dV$. This substitution, however, changes the contour of integration from that shown in Fig. 1.1 to contour C_2 in Fig. 1.2. Hence

$$C_K = \frac{Q_0}{\pi/D_K^*}\exp\left(\frac{\alpha x}{2} - \frac{\alpha^2 D_K^* t}{4}\right) I \qquad 1.19$$

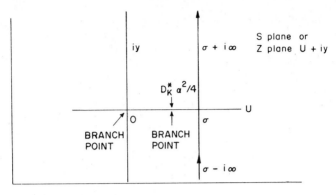

Figure 1.1 Bromich contour.

where

$$I = \int_{C_2} \frac{V \exp -[V^2 t + iVx/\sqrt{D_K^*}]}{\left(V - \dfrac{i\alpha\sqrt{D_K^*}}{2}\right)} dV$$

$$= \int_{C_2} f(V)\, dV \qquad\qquad 1.20$$

Cauchy's Residue theorem states (see Fig. 1.2)

$$\int_{C_2} f(V)\, dV + \int_{C_1} f(V)\, dV + \int_{C_3} f(V)\, dV + \int_{C_4} f(V)\, dV$$
$$= 2\pi i \sum \text{Res}\, (f(V)) \qquad 1.21$$

For $\alpha \gtrless 0$, there are no singularities inside the contour $C_1 + C_2 + C_3 + C_4$.

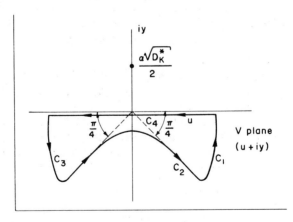

Figure 1.2 Contour C_2, Eq. 1.19.

Thus when $\alpha \geq 0$, $\sum$ Res $(f(V)) = 0$. However, if $\alpha < 0$, there is a simple pole at $V = i\alpha\sqrt{D_K^*}/2$ that contributes to the residue. One will, therefore, get a different answer depending on whether α is positive or negative. In either case, if one analyses $f(V)$ by letting $V = \rho e^{i\theta}$ for C_1 and C_3, $-\pi/4 < \theta < 0$, and $-\pi < \theta < -3\pi/4$, respectively, the term $\exp(-V^2 t)$ forces $f(V) \, dV \to 0$ as $\rho \to \infty$ on both C_1 and C_3. Hence

$$\int_{C_1} f(V) \, dV = \int_{C_3} f(V) \, dV = 0$$

Equation 1.21 therefore becomes, for $\rho \to \infty$,

$$I = \int_{C_2} f(V) \, dV = 2\pi i \, \text{Res} \, \frac{i\alpha\sqrt{D_K^*}}{2} - \int_{C_4} f(V) \, dV \qquad 1.22$$

$$= 2\pi i \, \text{Res} \, \frac{i\alpha\sqrt{D_K^*}}{2} + \int_{-\infty}^{\infty} f(V) \, dV$$

For $\alpha \geq 0$, it is understood that Res $(-i|\alpha|\sqrt{D_K^*}/2)$ is zero because the function $f(V)$ is analytic at the point $V = -i|\alpha|\sqrt{D_K^*}/2$. Thus

$$\int_{-\infty}^{\infty} f(V) \, dV = I$$

$$= \sqrt{D_K^*} \int_{-\infty}^{\infty} \frac{(2i\delta_0 - \alpha) \exp(-\delta_0^2 D_K^* t - i\delta_0 x)\delta_0 \, d\delta_0}{\delta_0^2 + \alpha^2/4}$$

$$1.23$$

with the substitution that $\delta_0 = V\sqrt{D_K^*}$. Hence one can write

$$I = 2\sqrt{D_K^*} \int_0^{\infty} \frac{\delta_0 \exp(-\delta_0^2 D_K^* t)[2\delta_0 \cos(x\,\delta_0) + \alpha \sin(x\,\delta_0)] \, d\delta_0}{\delta_0^2 + \alpha^2/4}$$

$$1.24$$

On the other hand, when $\alpha < 0$,

$$\int_{-\infty}^{\infty} f(V) \, dV = 2\sqrt{D_K^*} \int_0^{\infty} \frac{\delta_0 \exp(-\delta_0^2 D_K^* t)[2\delta_0 \cos(x\,\delta_0) + \alpha \sin(x\,\delta_0)] \, d\delta_0}{\delta_0^2 + \alpha^2/4}$$

$$1.25$$

in the same manner as that for Eq. 1.24. In addition, the residue is nonzero in that instance.

$$\text{Res}\left(-i|\alpha|\frac{\sqrt{D_K^*}}{2}\right) = -i|\alpha|\frac{\sqrt{D_K^*}}{2} \exp\left(\frac{\alpha^2 D_K^* t}{4} - \frac{|\alpha| x}{2}\right) \qquad 1.26$$

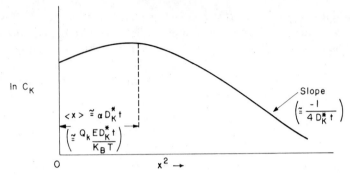

Figure 1.3 Diffusion with an applied electric field.

Therefore, combining Eq. 1.19 and 1.24 for $\alpha > 0$, one finds

$$C_K = \frac{Q_0}{\pi} \exp\left(\frac{\alpha x}{2} - \frac{\alpha^2 D_K^* t}{4}\right) \int_0^\infty \frac{\delta_0 e^{-\delta_0{}^2 D_K{}^* t}[2\delta_0 \cos(x\,\delta_0) + \alpha \sin(x\,\delta_0)]\,d\delta_0}{\delta_0{}^2 + \alpha^2/4}$$

$$1.27$$

When $\alpha < 0$

$$C_K = Q_0\left\{-\alpha \exp(\alpha x) + \frac{\exp\left(\dfrac{\alpha x}{2} - \dfrac{\alpha^2 D_K^* t}{4}\right)}{\pi}\right.$$

$$\left. \times \int_0^\infty \frac{\delta_0 e^{-\delta_0{}^2 D_K{}^* t}[2\delta_0 \cos(x\,\delta_0) + \alpha \sin(x\,\delta_0)]}{\delta_0{}^2 + \alpha^2/4}\,d\delta_0\right\} \quad 1.28$$

Figure 1.3 gives a plot of Eq. 1.27 for a hypothetical diffusion time with an applied field, E, in the positive x direction. For comparison, Fig. 1.4

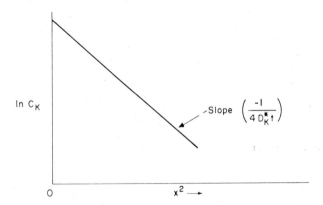

Figure 1.4 Diffusion in the absence of an applied field.

presents the result when $\alpha = 0$ ($E = 0$). One may set $\alpha = 0$ in Eqs. 1.27 and 1.28. Although with α nonzero one cannot express the answer, Eqs. 1.27 or 1.28, in a simple analytic form, when $\alpha = 0$ one finds a relatively simple answer. When $\alpha = 0$, Eqs. 1.27 and 1.28 become

$$C_K = \frac{2Q_0}{\pi} \int_0^\infty e^{-\delta_0^2 D_K^* t} \cos(x\,\delta_0)\,d\delta_0 = \frac{Q_0}{\sqrt{\pi D_K^* t}} \exp\left(\frac{-x^2}{4D_K^* t}\right) \qquad 1.29$$

Figure 1.4 gives a plot of Eq. 1.29. It is seen that the slope of ln C_K versus x^2 is $-1/(4D_K^* t)$. This is very convenient for an experimental determination of the diffusivity. One may vapor deposit some tracer isotope on the end of a long diffusion sample and anneal it at an elevated temperature for some period of time t. The diffusion coefficient in absence of any applied field is found by sectioning the sample and counting the decay of the isotope. On dividing the count rate by the section mass (proportioned to the volume), one plots the count rate/mass versus x^2. The slope is $-1/(4D_K^* t)$. Since t is known, one determines D_K^*.

Distinctly different results are achieved when the electric field is applied to the specimen. In that instance, the maximum concentration is not found at the origin but at some distance in the specimen $\cong (V_L + V_{FK})t$. A net drift of matter has been the result of the applied force. This type of result is typical of a force, such as an electric field or temperature gradient.

Finally, regarding Figs. 1.3 and 1.4, it is noteworthy that the concentration in Eq. 1.29 decreases by a factor of e when x^2 increases from 0 to $4D_K^* t$. Therefore, the concentration will fall to a negligible amount when $x \gg 2\sqrt{D_K^* t}$. This is a practical limit on the length of a finite diffusion specimen for which Eq. 1.29 is still applicable. Also, section thickness of the sample to determine data for the plot of ln C_K versus x^2 are inferred from this result. One wants to plot several data points each time the concentration falls by a factor of $1/e$. If, for example, five data points were desired for such a change in concentration, the section thicknesses would be $\approx 0.4\sqrt{D_K^* t}$. One knows how thick the sections will be from equipment limitations, and a diffusion time t must be sought to make the graph in Fig. 1.4 comparable by matching the equipment limiting section thickness to $0.4\sqrt{D_K^* t}$.

The LaPlace transform is not the only integral transform that is useful in the solution of diffusion problems. Fourier transforms also provide a convenient method of analysis. Whereas with the LaPlace method one transforms the time variable, the Fourier method is used to transform the spatial variable x. The next two examples show the technique of solution with Fourier transforms.

First, we resolve Eq. 1.10 with a set of boundary conditions that are comparable to the one used previously with the condition that D_K^* is constant for simplicity; the expected result is therefore Eq. 1.29 when $\alpha = 0$. The boundary conditions for this case are the following:

$$C_K(x, 0) = 2Q_0\, \delta(x)$$

$$C_K(x = \pm\infty, t) = \frac{\partial C_K}{\partial x}(x = \pm\infty, t) = 0$$

where $\delta(x)$ is the Dirac delta function. The factor of 2 is present because the sample is doubly infinite in this case; there is now twice the mass between $-\infty$ and $+\infty$ in comparison to 0 to ∞ before. Without the factor of 2, we would get one-half the previous answer for $\alpha = 0$.

The complex Fourier transform pair is

$$\phi(p, t) = \int_{-\infty}^{\infty} e^{-ipx} C_K(x, t)\, dx \qquad\qquad 1.30a$$

$$C_K(x, t) = \frac{1}{2\pi} \int_{-\infty}^{\infty} e^{ipx} \phi(p, t)\, dp \qquad\qquad 1.30b$$

Applying Eq. 1.30a to Eq. 1.10, one has

$$D_K^*\alpha \int_{-\infty}^{\infty} \frac{\partial C_K}{\partial x} e^{-ipx}\, dx = D_K^*\alpha \left\{ e^{-ipx} C_K \Big|_{-\infty}^{\infty} + ip\phi \right\} = i\alpha D_K^* p\phi$$

$$\int_{-\infty}^{\infty} D_K^* e^{-ipx} \frac{\partial^2 C_K}{\partial x}\, dx = D_K^* \left\{ e^{-ipx} \frac{\partial C_K}{\partial x} \Big|_{-\infty}^{\infty} + ip C_K e^{-ipx} \Big|_{-\infty}^{\infty} - p^2 \phi(p, t) \right\}$$

$$= -p^2 D_K^* \phi(p, t)$$

The last step results from the last boundary conditions; notice the integration by parts twice to transform the second derivative in x. Hence one has

$$\frac{d\phi(p, t)}{dt} = -D_K^* \phi(p^2 + i\alpha p)$$

which has a solution of the form

$$\phi(p, t) = A\, \exp\left[-D_K^* t(p^2 + i\alpha p)\right] \qquad\qquad 1.31$$

The transform of the first boundary condition is

$$\phi(p, 0) = \int_{-\infty}^{\infty} 2Q_0 \delta(x) e^{-ipx}\, dx = 2Q_0 \qquad\qquad 1.32$$

On comparing Eq. 1.31 and 1.32, one must include that

$$A = 2Q_0$$

Consequently,

$$C_K(x, t) = \frac{Q_0}{\pi} \int_{-\infty}^{\infty} \exp\left(ipx - p^2 D_K^* t - i\alpha p D_K^* t\right) dp \qquad 1.33$$

With the substitution $V = p\sqrt{(D_K^* t)}$, Eq. 1.33 becomes

$$C_K(x, t) = \frac{Q_0}{\pi\sqrt{(D_K^* t)}} \int_{-\infty}^{\infty} \exp\left(\frac{iVx}{\sqrt{(D_K^* t)}} - V^2 - i\alpha V\sqrt{(D_K^* t)}\right) dV$$

and the further substitution that

$$U = V - \frac{i}{2}\left[\frac{x}{\sqrt{(D_K^* t)}} - \alpha\sqrt{(D_L^* t)}\right]$$

further reduces the equation to

$$C_K(x, t) = \frac{Q_0 \exp -\frac{1}{4}[x/\sqrt{(D_K^* t)} - \alpha\sqrt{(D_K^* t)}]^2}{\pi\sqrt{(D_K^* t)}} \int_{-\infty}^{\infty} \exp\left(-u^2\right) du \qquad 1.34$$

$$= \frac{Q_0}{\sqrt{\pi D_K^* t}} \exp\left[-\frac{\{x - \alpha D_K^* t\}^2}{4 D_K^*}\right]$$

Since the last integral is equal to $\sqrt{\pi}$, Eqs. 1.34 and 1.29 are identical when $\alpha = 0$.

It is sometimes convenient to use two integral transforms in the solution of diffusion problems with constant coefficients. Again, if α is zero and D_K^* is constant in Eq. 1.10, the use of integral transforms is a possibility. The choice of transform depends on the initial and boundary conditions. The diffusion sample in the problem presented below is assumed to be semi-infinite in length, as before. The appropriate boundary conditions are the following:

$$C_K(0, t) = C_0$$

$$C_K(X, 0) = 0$$

$$\lim_{X \to \infty} C_K(X, t) = 0$$

$$\lim_{X \to \infty} \frac{dC_K}{dX} = 0$$

Since the first boundary condition gives constant concentration at the origin, the semi-infinite Fourier sine transform is appropriate; the last two conditions are also a necessity in the use of this transform, as will be seen. Furthermore, the second condition is a reasonable one when the Laplace transform is used. Consequently, this illustration shows the combination of both transforms.

We define the semi-infinite Fourier sine transform pair as follows:

$$\Phi = \int_0^\infty C_K \sin(\gamma x)\, dx \qquad\qquad 1.35a$$

$$C_K = \frac{2}{\pi} \int_0^\infty \Phi \sin(x\gamma)\, d\gamma \qquad\qquad 1.35b$$

Then,

$$\int_0^\infty \frac{d^2 C_K}{dx^2} \sin(\gamma x)\, dx = \left[\sin(\gamma x) \frac{dC_K}{dx} \bigg|_0^\infty - \gamma C_K \cos(\gamma x) \bigg|_0^\infty - \gamma^2 \Phi \right]$$

$$1.36$$

when one integrates by parts twice. The differential equation 1.10, with $\alpha = 0$, and with the first and last boundary conditions, becomes

$$\frac{1}{D_K^*} \frac{d\Phi}{dt} = -\gamma^2 \Phi + \gamma C_0 \qquad\qquad 1.37$$

The form of the Laplace transform pair is then

$$\bar{\Phi}(S) = \int_0^\infty e^{-st} \Phi(t)\, dt \qquad\qquad 1.38a$$

$$\Phi = \frac{1}{2\pi i} \int_{\sigma - i\infty}^{\sigma + i\infty} e^{ts} \bar{\Phi}(S)\, dS \qquad\qquad 1.38b$$

Applying Eq. 1.38a to Eq. 1.37, using a form analogous to Eq. 1.12 [the initial condition of $C_K(X, 0) = 0$], one finds

$$\bar{\Phi} = \frac{D_K^* \gamma C_0 / S}{D_K^* \gamma^2 + S}$$

From this and Eqs. 1.38b and 1.35b, one has

$$C_K = \frac{2}{\pi} \int_0^\infty \sin(x\gamma) \int_{\sigma - i\infty}^{\sigma + i\infty} C_0 \frac{D_K^*(\gamma/S) e^{st}\, dS\, d\gamma}{2\pi i (S + \gamma^2 D_K^*)} \qquad\qquad 1.39$$

In the complex integral over S, there is a simple pole at $S = 0$ and at $S = -\gamma^2 D_K^*$, and there are no other singularities in the function. Using the calculus of residues is therefore quite easy for the integral over S. One finds that

$$\int_{\sigma - i\infty}^{\sigma + i\infty} \frac{C_0 D_K^* \gamma / S e^{st} \, dS}{2\pi i (S + \gamma^2 D_K^*)} = 2\pi i \sum \text{Res} \, (0, -\gamma^2 D_K^*)$$

$$= C_0 \frac{1 - e^{-\gamma^2 D_K'}}{\gamma} \qquad 1.40$$

Substituting Eq. 1.40 into Eq. 1.39,

$$C_K = \frac{2 C_0}{\pi} \int_0^\infty \sin x\gamma (1 - e^{-\gamma^2 D_K^* t}) \frac{d\gamma}{\gamma} \qquad 1.41$$

The substitutions $\eta = x/\sqrt{D_K^* t}$ and $\mu = \sqrt{(D_K^* t)}$ are appropriate and alter Eq. 1.41 into

$$C_K = \frac{2 C_0}{\pi} \int_0^\infty \sin \mu\eta (1 - e^{-\mu^2}) \frac{d\mu}{\mu}$$

$$= C_0 \left\{ 1 - \frac{2}{\pi} \int_0^\infty \sin \mu\eta e^{-\mu^2} \frac{d\mu}{\mu} \right\}^{\cdot}$$

$$= C_0 \left[1 - \frac{2}{\pi} \int_0^\eta d\beta \left(\int_0^\infty e^{-\mu^2} \cos \mu\beta \, d\mu \right) \right]$$

$$= C_0 \left[1 - \frac{2}{\pi} \int_0^\eta d\beta \left(\frac{\sqrt{\pi}}{2} e^{-\beta^2/4} \right) \right]$$

$$= C_0 \left[1 - \text{erf} \left(\frac{\eta}{2} \right) \right] \qquad 1.42$$

where erf $(\eta/2)$ is defined by the integral

$$\text{erf } z = \frac{2}{\sqrt{\pi}} \int_0^z e^{-\beta^2} \, d\beta$$

and is a well-tabulated function. Figure 1.5 gives a graphical illustration of the form of Eq. 1.42.

Experimentally, this particular solution may be obtained by welding two long samples of a binary alloy together, each of which has a different concentration of component K. Then if the diffusivity is constant, α is zero, there is no volume change on mixing, and the solute concentration, C_K will follow

$$\frac{C_K - C_1}{C_2 - C_1} = 1 - \text{erf} \left(\frac{x}{2\sqrt{(D_K^* t)}} \right) \qquad 1.43$$

where $C_2 > C_1$ and C_2 is the initial solute concentration in one half of the welded specimen and C_1 is that in the other.

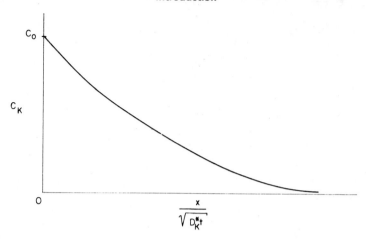

c_O

c_K

0

$$\frac{x}{\sqrt{D_K^* t}}$$

Figure 1.5 Concentration of semi-infinite slab with constant surface concentration.

The three previous problems illustrate the most common techniques of calculating concentration for the situation in which the diffusion coefficient and velocity per unit force are constant. They are the Laplace transform, the complex Fourier transform, and the infinite Fourier sine transform. The results give equations which, when compared to experimental data, yield information on the diffusivity and velocity from an applied force for the particular system for which the data was obtained.

Often, however, the velocity from an applied force, αD_K^*, is not constant. In fact, as shown later for a binary alloy, this term is often proportional to the gradient of the logarithm of the concentration. For a binary alloy one may often take Eq. 1.6 for substitutional alloys in the form

$$J_K = -D_K \frac{\partial C_K}{\partial x} \qquad\qquad 1.44$$

where $D_K = D_K^* r_K \alpha_K$. The term α_K, which is different from α, is the thermodynamic factor $(1 + d \ln \gamma_K / d \ln N_K)$ where γ_K is the chemical activity coefficient and N_K is the mole fraction of component K. Also, r_K is a vacancy wind factor when the solute is substitutional. Both α_K and r_K may depend on concentration and, furthermore, the lattice planes may be moving. In the coordinate referred to the lattice planes,

$$J_K = -D_K^* \frac{\partial C_K}{\partial x} + C_K V_{FK} \qquad\qquad 1.6$$

On comparing Eqs. 1.44 and 1.6

$$C_K V_{FK} = D_K^*(1 - r_K \alpha_K)\frac{\partial C_K}{\partial x} \qquad 1.45$$

Relative to a coordinate system fixed in space, for which Eq. 1.7 is applicable, one must introduce the velocity of the lattice, V_L. The flux in the fixed coordinate system is, therefore,

$$J_K^L = -D_K^*\frac{\partial C_K}{\partial x} + C_K(V_{FK} + V_L)$$

$$= -D_m\frac{\partial C_K}{\partial x} \qquad 1.46$$

The velocity of the lattice is related to the motion of both components of the alloy, and each component moves relative to the lattice following an equation similar to 1.44 or 1.6 with 1.45. Now both r_K and α_K are concentration dependent; therefore, D_m in Eq. 1.46 is concentration dependent. In Eq. 1.46 D_m is the mutual diffusivity because it is valid for both components (it is assumed that the volume change on mixing is negligible in Eq. 1.46). Substituting Eq. 1.46 into Eq. 1.7, one has

$$\frac{\partial C_K}{\partial t} = \frac{\partial}{\partial x}D_m\frac{\partial C_K}{\partial x} \qquad 1.47$$

Here, the complication is a variable diffusion coefficient, D_m. Integral transform techniques are not applicable to such a differential equation. The key to the solution of this differential equation is the coordinate change first introduced by Boltzmann; one lets a variable $\eta = x/\sqrt{t}$. Then, one finds that

$$\frac{\partial C_K}{\partial t} = \frac{dC_K}{d\eta}\frac{d\eta}{dt} = -\frac{1}{2}\frac{x}{t^{3/2}}\frac{dC_K}{d\eta}$$

$$\frac{dC_K}{dx} = \frac{dC_K}{d\eta}\frac{d\eta}{dx} = \frac{1}{t^{1/2}}\frac{dC_K}{d\eta}$$

and Eq. 1.47 becomes

$$\frac{-\eta}{2}\frac{dC_K}{d\eta} = \frac{d}{d\eta}\left(D_m\frac{dC_K}{d\eta}\right) \qquad 1.48$$

which is an ordinary differential equation. It is important to have initial conditions that are consistent with this transformation. If the concentration follows

$$C_K = C_1 \quad \text{for} \quad x > 0 \text{ at } t = 0$$

$$C_K = C_2 \quad \text{for} \quad x < 0 \text{ at } 0,$$

then the diffusion equation, 1.48, would yield Eq. 1.43 as a solution in the instance that $D_m = D = \text{constant}$. It is not expected that D_m be constant because of the thermodynamic factors: hence something else must be done. First, a pseudo-concentration, ψ, can be introduced as the dependent variable. On substituting

$$\psi = \frac{C_K - C_1}{C_2 - C_1}$$

into Eq. 1.48, one finds the equation

$$\frac{-\eta}{2}\frac{d\psi}{d\eta} = \frac{d}{d\eta}\left[D_m\frac{d\psi}{d\eta}\right] \qquad 1.49$$

which satisfies the conditions

$$\psi = 0 \quad \text{for} \quad x > 0 \text{ at } t = 0$$

and

$$\psi = 1 \quad \text{for} \quad x < 0 \text{ at } t = 0$$

Furthermore, since D_m is a function of C_K, it is also a function of ψ that the new differential equation satisfies the conditions

$$\psi = 0 \text{ at } \eta = \infty$$

and

$$\psi = 1 \text{ at } \eta = -\infty \qquad 1.50$$

The method of solution of Eq. 1.49 with conditions given by Eq. 1.50 is from Matano.[1] Note that one can write Eq. 1.49 as

$$\frac{-\eta}{2}d\psi = d\left[D_m\frac{d\psi}{d\eta}\right]$$

and the term on the right is an exact differential. Hence one has

$$-\int_{\psi=0}^{\psi=\psi'}\frac{\eta}{2}d\psi = \left[D_m\frac{d\psi}{d\eta}\right]_{\psi=0}^{\psi=\psi'}$$

From Fig. 1.6 it is seen that $d\psi/d\eta \to 0$ as $\psi \to 0$, therefore,

$$-\int_{\psi=0}^{\psi=\psi'}\frac{\eta}{2}d\psi = D_m\frac{d\psi}{d\eta}\bigg|_{\psi=\psi'} \qquad 1.51$$

The diffusion specimen is annealed at an elevated temperature for some time t at which point it is cooled to room temperature. At that time it is sectioned and one can determine and plot ψ versus $X/t^{1/2}$. Since the

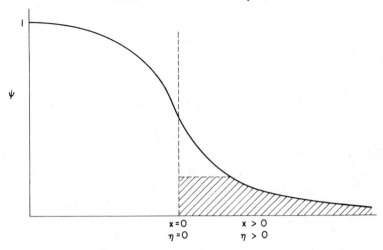

Figure 1.6 Matano integration of the diffusion equation.

time variable is subsequently fixed at some value t, the plot is proportional to ψ varying with X. Hence the reverse transformation is meaningful. That is,

$$D_m(\psi') = -\frac{1}{2t} \int_0^{\psi'} \frac{x\, d\psi}{\left(\dfrac{d\psi}{dx}\right)_{\psi'}} \qquad 1.52$$

The origin, $x = 0$, is found from the equation that follows:

$$\int_0^1 x\, d\psi = 0 \qquad 1.53$$

Finally, Eqs. 1.52 and 1.53 will determine the value of the variable diffusion coefficient at the concentration $\psi = \psi^1$.

Equations 1.52 and 1.53 present probably the most practical method of solution to Eq. 1.49. Alternative methods exist; however, they always present mathematical difficulties. As an example of the difficulties one would encounter, consider that the diffusivity can be reasonably represented by a series expansion in the concentration variable ψ. Then

$$D_m = D_m(\psi_0) + \frac{\partial D_m}{\partial \psi}\bigg|_{\psi_0} (\psi - \psi_0) + \frac{1}{2}\frac{\partial^2 D_m}{\partial \psi^2}(\psi - \psi_0)^2 + \cdots \qquad 1.54$$

If only the first two terms of Eq. 1.50 are used as a substitution in Eq.

1.49, one finds

$$\frac{-\eta}{2}\frac{d\psi}{d\eta}=\frac{d}{d\eta}\left\{\left[D_m(\psi_0)+\left.\frac{\partial D_m}{\partial \psi}\right|_{\psi_0}(\psi-\psi_0)\right]\frac{d\psi}{d\eta}\right\}$$

$$=D_m(\psi_0)\frac{d^2\psi}{d\eta^2}+\left.\frac{\partial D_m}{\partial \psi}\right|_{\psi_0}\left(\frac{d\psi}{d\eta}\right)^2 \qquad 1.55$$

Equation 1.55 is a nonlinear differential equation; hence it would be very difficult to solve. This is independent of the fact that the series expansion for the diffusivity is only accurate to the first two terms in the series; higher order expansions only present greater complications.

The previous examples have all dealt with infinite length samples. In practice, one closely approximates this condition whenever the length of the specimen, ℓ, is much greater than $2\sqrt{(D_K t)}$, as previously mentioned. Sometimes this is an impractical experimental limitation. In that instance, one must consider what would happen to the solution to the diffusion equation.

DIFFUSION IN A FINITE CRYSTAL

To this end, consider the differential equation 1.47. Let us assume that the sample length is ℓ and that the diffusivity, D_m, is in this case constant. The initial and boundary conditions are the following:

$$C_K(x, t = 0) = f(x)$$
$$J_K(x = 0, \ell) = 0$$

or equivalently the latter condition may be stated

$$\left.\frac{\partial C_K}{\partial x}\right|_{\substack{x=\ell \\ x=0}} = 0$$

The last condition imposes the limitation that the matter diffusing in the sample is trapped: hence the flux at $x = 0$ and $x = \ell$ is zero.

Since the diffusivity is constant, it is reasonable to use the method of separation of variables to determine the concentration. Therein, one assumes that $C_K = T(t)\bar{X}(x)$ and substitutes it into the differential equation. Thus Eq. 1.47 becomes

$$\frac{1}{T}\frac{dT}{dD_m t}=\frac{1}{\bar{X}}\frac{d^2\bar{X}}{dx^2}=\text{constant} \qquad 1.56$$

If the constant is $-\lambda_n^2$,

$$T = \exp - (\lambda_n^2 D_m t),$$
$$\bar{X} = A_n \sin (\lambda_n x) + B_n \cos (\lambda_n x)$$

and

$$\frac{d\bar{X}}{dx} = A_n \lambda_n \cos (\lambda_n x) - B_n \lambda_n \sin (\lambda_n x)$$

Now

$$\frac{\partial C_K}{\partial x} = 0 \quad \text{if} \quad \frac{d\bar{X}}{dx} = 0$$

so that one has at the point $x = 0$, $A_n = 0$. At $x = \ell$, one can force $dX/dx = 0$ by letting $\lambda_n = n\pi/\ell$. Thus

$$C_K = \sum_{n=0}^{\infty} B_n \cos \left(\frac{n\pi x}{\ell}\right) \exp - \left[\left(\frac{n\pi}{\ell}\right)^2 D_m t\right] \qquad 1.57$$

When $t = 0$, the initial condition $C_K(x, t = 0) = f(x)$ may be applied using the orthogonality of the cosine functions. Hence for $t = 0$,

$$\int_0^\ell f(x) \cos \left(\frac{m\pi x}{\ell}\right) dx = \sum_{n=0}^{\infty} B_n \int_0^\ell \cos \left(\frac{n\pi x}{\ell}\right) \cos \left(\frac{m\pi x}{\ell}\right) dx$$

Therefore, it follows that

$$B_m = \frac{\displaystyle\int_0^\ell f(x) \cos \left(\frac{m\pi x}{\ell}\right) dx}{\displaystyle\int_0^\ell \cos^2 \left(\frac{m\pi x}{\ell}\right) dx}$$

$$= \frac{2}{\ell} \int_0^\ell f(x) \cos \left(\frac{m\pi x}{\ell}\right) dx \qquad 1.58$$

Combining Eqs. 1.57 and 1.58, the solution to the differential equation is, therefore,

$$C_K = \frac{2}{\ell} \sum_{n=0}^{\infty} \left[\int_0^\ell f(x) \cos \left(\frac{n\pi x}{\ell}\right) dx\right] \cos \left(\frac{n\pi x}{\ell}\right) \exp - \left[\left(\frac{n\pi}{\ell}\right)^2 D_m t\right]$$

$$1.59$$

Equation 1.59 is cumbersome to use in diffusion experiments because the series usually converges somewhat slowly. Furthermore, the determination of the diffusion coefficient from experimental data using this equation becomes difficult. For that matter, the determination of D_K^* and α from Eqs. 1.27 and 1.28 are equally as difficult. A closed form solution to the differential equation in terms of either elementary or tabulated functions is needed. Then the diffusion parameters D_K^*, α, or D_m would be readily determinable from data. Consequently, one normally tries to avoid experimental conditions leading to diffusion results like those found in Eqs. 1.27 and 1.28 for $\alpha \neq 0$ and in Eq. 1.59. Preferably, the experimentalist will attempt to have his conditions compatible with Eqs. 1.29, 1.34, 1.42, 1.43, or 1.52. This, however, is not always possible.

Before leaving this section on the mathematics of diffusion in continua, mention should be made of the complications that arise when dealing with heterogeneous or multiphase media. Such problems exist both in fundamental research and in applications. It is always true that the conservation equation 1.7 will hold in a homogeneous media at rest relative to a fixed coordinate system. If one has the growth of phases to consider, the boundary of the phase will be moving, and provision must be accounted for in such a problem because Eq. 1.7 will no longer be applicable. An equation similar to Eq. 1.7 for a moving coordinate system must be devised. Such an equation will be applicable in each homogeneous region of the material. At phase boundaries, compatibility conditions to link the concentration of a component in one phase to the concentration in another will be necessary. The second law of thermodynamics leads one to believe that such a compatibility condition is the continuity of the chemical potential across the phase boundary. A second condition is often needed, also, and that is provided by the continuity of the flux of the component in question across the phase boundary. Thus the concentration of component K and its gradient are not normally expected to be continuous across phase boundaries of heterogeneous media. The obvious example of such a diffusion problem would be oxidation of some metal where two or more oxides of different stoichiometry are being formed.

INFLUENCE OF NEUMANN'S PRINCIPLE ON THE DIFFUSIVITY TENSOR

The symmetry of the crystalline lattice into which one is conducting a diffusion experiment influences the components of the diffusivity tensor. It may be argued that the components of the diffusivity are restricted by what is called Neumann's principle. Let the coordinate axes be x_1, x_2, and x_3 in this section. The flux of component K in the absence of applied force may thus be written

$$J_i = - \sum_{j=1}^{3} D_{ij} \frac{\partial C_K}{\partial x_j} \qquad 1.60$$

where the subscript i is the component of the flux in the direction x_i.

Neumann's principle asserts that if a coordinate transformation follows the symmetry relation of the crystalline lattice in question, then that transformation leaves the diffusivity tensor invariant. A general tensor coordinate transformation for the diffusivity from the coordinate system x_i into a new coordinate system $\bar{x}_i$ is accomplished in the following manner. Let D'_{ij} represent the diffusivity tensor component in the $\bar{x}_i$ coordinate system. Then

$$D'_{ij} = \sum_{K=1}^{3} \sum_{\ell=1}^{3} \frac{\partial \bar{x}_i}{\partial x_K} \frac{\partial \bar{x}_j}{\partial x_\ell} D_{K\ell}$$

$$= \sum_{K=1}^{3} \sum_{\ell=1}^{3} a_{iK} a_{j\ell} D_{K\ell} \qquad\qquad 1.61$$

when $\partial \bar{x}_i/\partial x_K = a_{iK}$, the direction cosine for the transformation.

Neumann's principle asserts that if a_{iK} follows a crystalline symmetry operation, then $D'_{ij} = D_{ij}$.

Examples of such rotations are given in the next three equations:

1. Twofold axis of rotation coincident with the x_1 axis, the matrix (a) with elements a_{ij} is found as

$$(a) = \begin{bmatrix} 1 & 0 & 0 \\ 0 & -1 & 0 \\ 0 & 0 & -1 \end{bmatrix}$$

2. Threefold axis of rotation coincident with the x_1 axis

$$(a) = \begin{bmatrix} 1 & 0 & 0 \\ 0 & -\dfrac{1}{2} & \dfrac{\sqrt{3}}{2} \\ 0 & -\dfrac{\sqrt{3}}{2} & -\dfrac{1}{2} \end{bmatrix}$$

3. Fourfold axis of symmetry coincident with the x_i axis of rotation

$$(a) = \begin{bmatrix} 1 & 0 & 0 \\ 0 & 0 & 1 \\ 0 & -1 & 0 \end{bmatrix}$$

When transformations of this nature are applied to crystals, the diffusion coefficient written as a matrix has the following form:

$$(D) = \begin{pmatrix} D_{11} & 0 & 0 \\ 0 & D_{11} & 0 \\ 0 & 0 & D_{11} \end{pmatrix}$$

for a cubic crystal;

$$(D) = \begin{pmatrix} D_{11} & 0 & 0 \\ 0 & D_{11} & 0 \\ 0 & 0 & D_{33} \end{pmatrix}$$

for a hexagonal crystal, and so forth.

To give an illustration of the procedure followed in this type of analysis, consider the transformations necessary for a two- instead of a three-dimensional space, as given in Eq. 1.62. Then, one has

$$D'_{ij} = \sum_{K=1}^{2} \sum_{\ell=1}^{2} a_{iK} a_{j\ell} D_{K\ell} \qquad\qquad 1.62$$

Next consider what symmetry would exist for a two-dimensional hexagonal array of lattice points. Expanding Eq. 1.62 one finds

$$D'_{11} = a_{11}a_{11}D_{11} + a_{11}a_{12}D_{12} + a_{12}a_{11}D_{21} + a_{12}a_{12}D_{22} \qquad 1.63a$$
$$D'_{12} = a_{11}a_{21}D_{11} + a_{11}a_{22}D_{12} + a_{12}a_{21}D_{21} + a_{12}a_{22}D_{22} \qquad 1.63b$$
$$D'_{21} = a_{21}a_{11}D_{11} + a_{21}a_{12}D_{12} + a_{22}a_{11}D_{21} + a_{22}a_{12}D_{22} \qquad 1.63c$$
$$D'_{22} = a_{21}a_{21}D_{11} + a_{21}a_{22}D_{12} + a_{22}a_{21}D_{21} + a_{22}a_{22}D_{22} \qquad 1.63d$$

The following two transformations show mirror symmetry and threefold symmetry applicable to a two-dimensional hexagonal array of points.

$$(a) = \begin{pmatrix} -1 & 0 \\ 0 & 1 \end{pmatrix} \qquad\qquad 1.64$$

$$(a) = \begin{pmatrix} -\dfrac{1}{2} & \dfrac{\sqrt{3}}{2} \\ -\dfrac{\sqrt{3}}{2} & -\dfrac{1}{2} \end{pmatrix} \qquad\qquad 1.65$$

On applying Eq. 1.64 to Eqs. 1.63a–d, one finds that $D'_{11} = D_{11}$, $D'_{12} = -D_{12}$, $D'_{21} = -D_{21}$, and $D'_{22} = D_{22}$. Since Eq. 1.64 is a symmetry transformation for the hexangonal array, $D'_{ij} = D_{ij}$ by Neumann's principle, and one concludes that $D_{12} = D_{21} = 0$ from the above. Furthermore, when one subsequently applies Eq. 1.65 to Eqs. 1.63$a-d$, one finds

$$D'_{11} = \tfrac{1}{4}D_{11} + \tfrac{3}{4}D_{22}$$
$$D'_{22} = \tfrac{3}{4}D_{11} + \tfrac{1}{4}D_{22}$$

1.66

Finally, since $D'_{ij} = D_{ij}$, Eq. 1.66 implies that $D_{11} = D_{22}$ for the hexagonal two-dimensional array. Thus for that lattice,

$$(D) = \begin{pmatrix} D_{11} & 0 \\ 0 & D_{11} \end{pmatrix}$$

1.67

As a consequence, it is apparent that one can select a particular crystallographic axis for diffusion experiments in single crystals and the diffusivity used in the above development can be considered a constant. Neumann's principle, therefore, reduces the diffusivity tensor to a scalar for most experimental work. Thus, for example, in a cubic lattice the diffusivity is a scalar independent of the orientation of the lattice; one obtains the same diffusivity in a $\langle 111 \rangle$ direction that one would find in a $\langle 100 \rangle$ direction. We find such considerations of great practical value to our discussion of the correlation factor in Chapter 3.

On the other hand, when one is studying diffusion in a crystal system where the diffusion coefficient is not isotropic, provision must be made for the orientation of the diffusion direction to the principal crystallographic axes. If for example, the concentration gradient is at some angle θ to the c crystallographic axis of a hexagonal single crystal, the diffusion coefficient is found to be

$$D(\theta) = D_{11} \sin^2 \theta + D_{33} \cos^2 \theta$$

1.68

The same equation would be valid for a tetragonal crystal where D_{33} is again the diffusivity component along the c axis.

REFERENCE

1. C. Matano, *Japan. Phys.*, **8**, 109 (1933).

CHAPTER 2

DIFFUSION OF INTERSTITIAL SOLUTES

ATOMIC VIBRATIONS IN CRYSTALS

To calculate the diffusion coefficient of a solute in a crystal, one must be able to express the contribution of a solute atom and its neighbors to the kinetic energy of the crystal.

Consider such an atom residing in some crystal and presume that you attempt a calculation of that atom's contribution to the thermodynamic internal energy of the crystal. The simplest place to start with such a calculation is in the potential energy of the particular atom, not necessarily an interstitial as yet, assuming that all other atoms are at rest, sitting on their respective atom sites. If this particular atom, which we label i, is also at rest on its site, one may consider virtual displacements of the atom about its equilibrium position. Letting ϕ_i be the potential energy of the ith atom, one may take a series expansion considering the small displacements, δ_x, δ_y, and δ_z, of atom i. To second order, such a series expansion leads to the following expression

$$\phi_i(\delta_x, \delta_y, \delta_z) = \phi_i(0) + \tfrac{1}{2}F_i(\delta_x{}^2 + \delta_y{}^2 + \delta_z{}^2) + \cdots \qquad 2.1$$

where $\phi_i(0)$ is the binding energy of atom i to its site. Notice the absence of a term that is linear in δ_x, δ_y, and δ_z; there is no net force on atom i at $\delta_x = \delta_y = \delta_z = 0$. If the displacements are truly small, Eq. 2.1 is of sufficient accuracy to describe the energy. For such displacements, classical mechanics tells us that the kinetic energy of the particle will be that of a harmonic oscillator that vibrates at a frequency

$$\nu_i = \frac{1}{2\pi}\sqrt{\frac{F_i}{m_i}} \qquad 2.2$$

where m_i is the mass of atom i. Quantum mechanics, on the other hand, tells us the energy of the particle will be that of a quantized harmonic oscillator whose energy is ϵ_i.

$$\epsilon_i = \phi_i(0) + h\nu_i(n_x + n_y + n_z + \tfrac{3}{2}) \qquad 2.3$$

where $n_x = 0, 1, 2, \cdots$, and the same is true for n_y and n_z. The constant h is Planck's constant, and ν is the frequency found in Eq. 2.2. It is assumed here that this is the only contribution that atom i makes to the energy of the crystal. Thus any electronic interactions are visualized only as contributing to $\phi_i(0)$ and F_i. With such an assumption, one may calculate the contribution that atom i makes to the Helmholtz free energy of the crystal. At low pressures, on the order of 1 atm, the Gibbs' and Helmholtz free energies are nearly equal for condensed phases, and we use them interchangeably as convenient. At elevated pressures, the atomic volume of atom i is altered and the binding potential, $\phi_i(0)$, and force constant, F_i, will be changed. Thus when one permits the pressure to be elevated, both ν_i and $\phi_i(0)$ are sensitive to changes in atomic volume. For that matter, the position of atom i in the crystal, whether it is an interstitial or substitutional atom, whether it is isolated from other atoms of its own chemical species or clustered, will influence the energy terms $\phi_i(0)$ and ν_i. That is, bonding effects will influence the energy.

In its present position, atom i will make a contribution to the potential and kinetic free energy of the crystal of magnitude A_i where A_i is found by statistical mechanics to be the following:

$$A_i = -K_B T \ln\left\{ \sum_{n_x=0}^{\infty} \sum_{n_y=0}^{\infty} \sum_{n_z=0}^{\infty} \exp\frac{-\epsilon_i}{K_B T}\right\} \qquad 2.4a$$

$$= \phi_i(0) - 3K_B T \ln g(\nu_i) \qquad 2.4b$$

where

$$g(\nu_i) = \frac{e^{-h\nu_i/2K_B T}}{1 - e^{-h\nu_i/K_B T}} \qquad 2.5$$

For the purposes of discussion of diffusion and atom jump frequencies, we are generally interested in the properties of the crystal at high temperatures. At these elevated temperatures, atom i contributes to the internal energy as magnitude E_i and vibrational entropy as magnitude S_i.

These may be found from A_i as follows: substitute Eq. 2.5 into Eq. 2.4 and take the limit where $T \gg h\nu_i/K_B$. Then Eq. 2.4 becomes

$$A_i = \phi_i(0) + \tfrac{3}{2}h\nu_i + 3K_BT \ln \frac{h\nu_i}{K_BT} \qquad 2.6$$

Now since

$$E_i = \frac{\partial\left(\dfrac{A_i}{T}\right)}{\partial\left(\dfrac{1}{T}\right)}\Bigg|_{\nu_i}$$

one finds

$$E_i = \phi_i(0) + \tfrac{3}{2}h\nu_i + 3K_BT \qquad 2.7a$$

and

$$S_i = 3K_B\left(1 - \ln \frac{h\nu_i}{K_BT}\right) \qquad 2.7b$$

Equation 2.7a leads directly to the well-known law of Dulong and Petit whereby the atomic contribution to the heat capacity is $3K_B$; the energy and entropy of the entire crystal is related to the sum of the contributions from each atom. For a single component system, one would find the internal energy E and entropy S is

$$E = \sum_i E_i \qquad 2.8a$$

$$S = \sum_i S_i \qquad 2.8b$$

For a multicomponent system Eq. 2.8 is an oversimplification due to configurational effects. In a calculation such as that given in Eq. 2.8, however, one must perform an analysis for the normal modes of vibration ν_i. The Debye approximation is a well-known analysis of this type whereby the distribution of vibrational frequencies is analyzed using a continuum approach. Since ν_i in Eq. 2.7 for a pure crystal would be the maximum frequency, shortest wavelength, the normal mode analysis gives the other frequencies of longer wavelength.

As a consequence of the foregoing model for the kinetic energy of the crystal, one must view each atom of a crystal vibrating about its equilibrium position with a frequency ν, which is the maximum vibrational frequency in the crystal. For a pure metal, this will be the so-called Debye frequency. Conceivably, for an interstitial solute of very small mass, Eq. 2.2 suggests that the maximum vibrational frequency could be larger than the Debye frequency of the solvent. As time proceeds, one can visualize that the atom is attempting to jump out of its site throughout each oscillation, and every now and then the waves contributing to the complex vibration could superimpose such that a jump to a neighboring empty but equivalent site† is possible. The number of times per second a particle of component K makes a successful jump to such a site is Γ_K, the jump frequency. This jump frequency will be the product of the maximum number of jump attempts per second that the atom makes, ν_K, times the probability that any single attempt is successful, P_K. Hence

$$\Gamma_K = \nu_K P_K \qquad 2.9$$

Since the atom is normally constrained to its equilibrium position, that constraint provides an energy barrier through which the atom must pass for the jump attempt to be a success. The term P_K is the probability that the atom can accomplish this fluctuation in energy. Such a fluctuation can be thought of in several ways. The atom could itself simply attain sufficient energy to surmount the forces from its neighbors such that it is free to move to an adjacent site. Equivalently, the neighboring atoms could attain sufficient energy to move out of the way of the jumping particle leaving it free to make its displacement. Most likely, there is probably some mixture of both processes that occurs on the average. One may view, therefore, a fluctuation in the kinetic energy of the diffusing atom and of the atoms that surround it. In the process of this motion, the potential energy of the particles involved is altered such that Eq. 2.1 is no longer applicable. Some other function must replace the approximation used in Eq. 2.1. If one makes a series expansion similar with Eq. 2.1 for that portion of the atom's trajectory where the maximum in energy occurs, one expects a change in $\phi_i(0)$ and f_i for each atom participating in the jump, the moving atom and its neighbors. Each of these atoms will experience a change in free energy given by (see Eq. 2.6).

$$\Delta A_i = \Delta \phi_i(0) + \tfrac{3}{2} h \, \Delta \nu_i + 3 K_B T \ln \frac{\nu_{if}}{\nu_{io}} \qquad 2.10$$

ν_{if} and ν_{io} are the frequencies associated with the activated and original states, respectively.

One may sum Eq. 2.10 over the atoms that participate in the fluctuation leading to a jump giving the Gibbs' free energy of activation ΔG_K for

---

† A neighboring but equivalent site is defined by the mechanism of diffusion that is occurring. For an interstitial mechanism, such a site is an adjacent nearest neighbor, and empty interstitial site. For a substitutional solute diffusing by a vacancy mechanism, the site must be a nearest neighbor vacant lattice site into which the solute could jump.

an atom of component K:

$$\Delta G_K = \sum_i \Delta A_i + \Delta(P\underline{V})_i \qquad 2.11$$

If pressure and volume changes are also included as indicated, then one must interpret ΔG_K in Eq. 2.11 as a Gibbs' free energy change. Statistical mechanics gives us the steady-state probability that a fluctuation in free energy is of magnitude ΔG_K. This is

$$P_K = \exp\left(\frac{-\Delta G_K}{K_B T}\right) \qquad 2.12$$

The frequency of atom jumps of component K will be, considering Eqs. 2.9 and 2.12,

$$\Gamma_K = \nu_K \exp\left(\frac{-\Delta G_K}{K_B T}\right) \qquad 2.13$$

JUMP FREQUENCY GRADIENT

Equation 2.13 might imply that the jump frequency is independent of direction. This, however, may not be true because one is concerned with a thermodynamic quantity in the exponential. The term $\Delta G_K/K_B T$ is a thermodynamic function that depends on certain intensive thermodynamic variables: pressure, temperature, and concentration. If any of these three variables is dependent on position within the crystal, for example, the coordinate axis x, then $\Delta G_K/K_B T$ is probably also dependent on the variable x. If the x projection of an atom jump between two neighboring sites is b, and if one assumes that the energy term $\Delta G_K/K_B T$ is acting at the midpoint of the atom's trajectory, $b/2$, then for a jump in the $+x$ direction, one has

$$\Gamma_{+K} = \Gamma_K(x=0) + \frac{b}{2}\frac{\partial \Gamma_K}{\partial x} = \Gamma_K(x=0)\{1 + \delta_K\} \qquad 2.14$$

where

$$\delta_K = \frac{b}{2\Gamma_K(x=0)}\frac{\partial \Gamma_K}{\partial x} = -\frac{b\partial \Delta G_K/K_B T}{2\partial x}$$

This gradient term is expected whenever a gradient in pressure, temperature, or concentration could give rise to δ_K being nonzero.

The jump in the negative x direction would also be gradient-dependent as follows:

$$\Gamma_{-K} = \Gamma_K(x=0) - \frac{b}{2}\frac{\partial \Gamma_K}{\partial x}$$

$$= \Gamma_K(x=0)\{1 - \delta_K\} \qquad 2.15$$

JUMP FREQUENCY RESPONSE TO AN APPLIED FIELD

Independent of the considerations leading to Eqs. 2.14 and 2.15, several applied fields can alter the jump characteristics of the moving atom. This response leads to the velocity per unit force term mentioned in Eq. 1.6. The jump frequency gradient term mentioned in the previous section, on the other hand, leads to the diffusion coefficient gradient term in Eq 1.9. These ideas are shown more clearly in this and a following section.

A chemical activity coefficient gradient, electric field gradient, a temperature gradient, and so forth, all act on the jumping atom as if they were exerting a force to bias the jump rate in the direction of the applied field. This force, independent of whichever of the fields is operative, causes the jump frequency change in the $+x$ direction which is related to an energy change $\Delta H = (b/2)F$ where $b/2$ is the distance over which the force, F, acts for the atom between its ground and activated state. This distance is projected along the x axis; the x axis is collinear with the force F. The action of this force is to change the jump probability as follows:

$$\Gamma_{+K} = \Gamma_K(F=0)\exp\left\{\frac{bF}{2K_BT}\right\}$$

$$= \Gamma_K(F=0)\left\{1 + \frac{bF}{2K_BT}\right\}$$

$$= \Gamma_K(F=0)\{1 + \epsilon_K\} \qquad 2.16$$

In the negative field direction, the force retards the jump in a similar manner to the above

$$\Gamma_{-K} = \Gamma_K(F=0)\{1 - \epsilon_K\} \qquad 2.17$$

Generally, when an arbitrary field acts on the diffusing atom, it could act to yield both responses, the jump frequency gradient, Eq. 2.14, and the applied field response, Eq. 2.16. For a pressure, chemical activity, or temperature gradient, such a situation exists and in that instance

$$\Gamma_{+K} = \Gamma_K\{1 + \delta_K + \epsilon_K\}$$

$$\Gamma_{-K} = \Gamma_K\{1 - \delta_K - \epsilon_K\} \qquad 2.18$$

An electric field for example does not influence the diffusion coefficient to any appreciable extent, and consequently, Eq. 2.18 for an electric field as the only force yields

$$\delta_K = 0 \qquad\qquad 2.19$$

However, the electric field does alter the jump frequency in a manner similar to that given in Eq. 2.16. Hence

$$\epsilon_K = \frac{Eqb}{2K_BT} \qquad\qquad 2.20$$

where E is the field intensity, q is the effective ionic charge of the jumping ion, and $b/2$ is the x projection of the distance to the activated state. Then, Eq. 2.18 becomes, $\delta_K = 0$,

$$\Gamma_{\pm K} = \Gamma_{0K}\{1 \pm \epsilon_K\} \qquad\qquad 2.21$$

with ϵ given in Eq. 2.20 in this instance.

The subscript 0 denotes the absence of any field. Equation 2.18 for a temperature gradient, however, will have two terms as follows:

$$\delta_K = \frac{b}{2} \frac{\partial \Delta G_K / K_B T}{\partial T} \frac{\partial T}{\partial x} \qquad\qquad 2.22a$$

$$\epsilon_K = \frac{-b}{2} \frac{q_K^*}{K_B T^2} \frac{\partial T}{\partial x} \qquad\qquad 2.22b$$

Where q_K^* is the heat of transport of the atomic jump of component K; it is the amount of energy that the diffusing atom carries when it moves the distance b in the direction of the field (temperature gradient). This energy contributes to the amount of heat transferred in a temperature gradient. Since both δ and ϵ in Eq. 2.22 are nonzero, there are two contributions to the jump frequency bias given by Eq. 2.18

$$\Gamma_{\pm K} = \Gamma_{0K}\{1 \pm \delta_K \pm \epsilon_K\} \qquad\qquad 2.23$$

with δ and ϵ given by Eqs. 2.22a and b, respectively.

Consequently, for the gradients in concentration, temperature, and pressure, the jump frequency is biased relative to the field with a result similar to Eq. 2.23. For the gradients of electric field, gravitational field, and so forth, there are no apparent terms of the types δ_K given in Eq. 2.14. Then, the frequency is biased in a manner similar with Eq. 2.21.

DIFFUSION MECHANISMS

The mechanism by which an atom may migrate during diffusion is to a great extent controlled by the atom's immediate surroundings. For example, in a gas one may explain not only diffusion but also such thermodynamic properties as internal energy and the equation of state by assuming the atoms move with a given momentum in a particular direction. It continues to do so until it suffers a collision. Such a set of events defines the average kinetic energy and mean free path for the moving atom, and this leads to a satisfactory kinetic theory.

For liquids and amorphous solids, the kinetic theory for diffusion and the thermodynamic properties is a very complicated and not a terribly satisfactorily defined subject. Theories of diffusion have treated the liquid either as a dense gas or as nearly a solid. Neither treatment has been entirely satisfactory for diffusion; however, the thermodynamic properties have best been explained using statistical mechanics peculiar to the liquid state. Diffusion will likely have to be treated similarly.

For solids the mechanism of diffusion is related to the manner in which the atom in question is found in solution in the particular system. An interstitial solute in a metal remains on interstitial sites for the most part, although it may occasionally annihilate a vacancy. Thus in the types of diffusion mechanisms, one must incorporate the interstitial mechanism, Therein, an atom on an interstitial site moves to an adjacent empty interstitial site. Figure 2.1 illustrates this process for the fcc lattice. The interstitial sublattice is also fcc in that instance. A common system where the interstitial mechanism is believed to occur is exemplified by the motion of gases in metals, for example, hydrogen moving in iron.

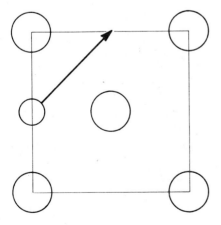

Figure 2.1 Interstitial diffusion mechanism in an fcc lattice; a (100) plane projection of the unit cell.

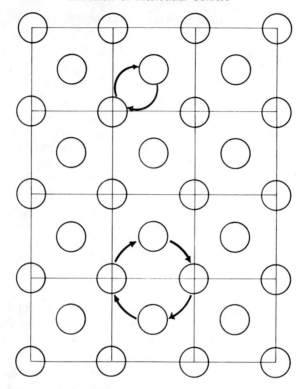

Figure 2.2   Two- and four-atom ring mechanism for the fcc lattice.

The motion of substitutional solutes becomes somewhat more complicated because in the majority of cases the solute occupies a lattice site before and after the atomic jump. Perhaps the simplest case for such diffusion are the so-called ring mechanisms. A ring mechanism is identified when a group of atoms jump simultaneously to neighboring sites. The configuration of the group of atoms is that of a ring, as illustrated in Fig. 2.2, where a two-atom and four-atom ring is given for the fcc crystal system. The two-atom ring mechanism (sometimes called the exchange mechanism) is not seriously considered for any close packed crystal, fcc, or hcp, because of the very large distortion to the lattice that would be associated with the jump. Such a distortion would require the collection of a great deal of energy, which is a rare event. The many-atom ring mechanism does not have this fault; however, the simultaneous motion of a large number of atoms is likely to be a rare event also. There the objection is the magnitude of the entropy and not the energy.

The vacancy mechanism may well be the most predominant mechanism for the motion of substitutional solutes in metals. In this mechanism, one relies on the fact that a certain fraction of lattice sites are empty at any temperature above absolute zero. Thus there is a small but finite probability that a neighboring site to a solute is vacant. Then, the solute may displace by jumping into the vacant lattice site. Figure 2.3 illustrates this concept again for the fcc lattice. Also shown on the figure is the divacancy mechanism. At temperatures approaching the melting point of most metals, the fraction of vacant lattice sites becomes sufficiently large that one cannot neglect the possibility of two adjacent vacant lattice sites. This

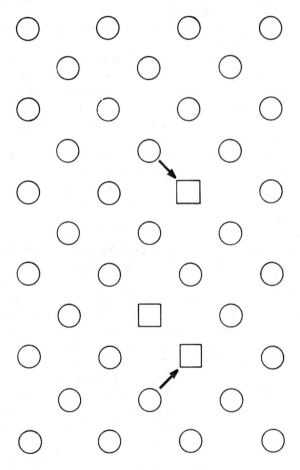

Figure 2.3 The vacancy and divacancy mechanism for diffusion in an fcc lattice. Divacancy dissociation is shown.

is particularly true since there is a tendency for these defects to bind together. As an example, in fcc there are 24 broken bonds around two isolated vacancies but only 18 around a divacancy. This causes two effects; first there is a tendency for the vacancies to bind together and second, they tend to jump at a faster rate due to the increased lattice distortion around a vacancy pair. A thorough discussion of the vacancy mechanism is given in Chapters 3 and 4. Divacancies are more complicated and are discussed in Chapters 5 and 6.

Other mechanisms that have been considered to be important for substitutional solutes are the interstitialcy and crowdion. The interstitialcy is the motion of an atom that is normally a substitutional as follows: The atom has arrived at an interstitial site by some means. It can then move through the lattice with a small distortion by shoving a neighboring substitutional atom into an interstitial site and acquiring the substitutional site itself. The interstitialcy mechanism is thought to occur in some salts, for example, AgBr. The crowdion is an extra ion along a particular close-packed direction in the lattice. The crowdion moves by displacing the extra atom along the close-packed direction. The crowdion is likely to be important during the anneal following radiation damage; a flux of high energy particles could create such a defect.

Most of the remainder of the chapter concerns itself with diffusion by the interstitial mechanism. For substitutional solutes, crystalline defects such as dislocations and grain boundaries are important, particularly at lower temperatures. These topics are reserved for Chapter 7.

DIFFUSION FLUX FOR AN INTERSTITIAL–RANDOM WALK CALCULATION

Consider three adjacent planes on which interstitial solutes can reside in some crystal. These planes are normal to a field applied in the positive x direction, and the crystal is oriented in a manner that the interstitial solute may jump between adjacent planes only. In this manner, if b is the spacing between planes, b is also the magnitude of the x projection of an interstitial jump. Hence for octahedral sites in fcc or bcc crystals, the planes are (100) and the field is applied in the $\langle 100 \rangle$ direction. Furthermore, the field, of unspecified origin at present, may cause a jump frequency gradient as well as a drift velocity per unit force as described in the previous two sections.

To calculate the diffusion flux of interstitial component K, consider the number of atoms per unit area of component K on plane i; $i = 1, 2$, or 3 as above. This number is N_i. Any of these atoms can jump to its neighboring plane at a jump rate $\Gamma_{ijK'}$ when this term is the rate of jumps of component K from plane i to plane j, i; $j = 1$, 2 or 3. The diffusion flux across plane 2 is taken one-half the difference in the rate at which atoms arrive and leave plane 2 in the positive and negative x directions. Thus

$$J_K = \tfrac{1}{2}\{N_1\Gamma_{12K} + N_2\Gamma_{23K} - N_2\Gamma_{21K} - N_3\Gamma_{32K}\} \qquad 2.24$$

The local concentration of component K on plane i, $C_{iK} = N_i/b$. These concentrations are related to first order in small quantities through a series expansion. Hence one may take

$$\begin{aligned}
C_{3K} &= C_{2K} + b\, \partial C_K/\partial x \\
C_{1K} &= C_{2K} - b\, \partial C_K/\partial x
\end{aligned} \qquad 2.25$$

The substitution of Eq. 2.25 into Eq. 2.24 yields

$$\begin{aligned}
J_K = \frac{b^2}{2}\Bigg\{ & \frac{N_1 - N_2}{b^2}\frac{\Gamma_{12K} + \Gamma_{21K}}{2} \\
& + \left[\frac{N_1 + N_2}{2b}\right]\frac{\Gamma_{12K} - \Gamma_{21K}}{b} \\
& + \left[\frac{N_2 - N_3}{b^2}\right]\frac{\Gamma_{23K} + \Gamma_{32K}}{2} \\
& + \left[\frac{N_2 + N_3}{2b}\right]\frac{\Gamma_{23K} - \Gamma_{32K}}{b} \Bigg\} \\[4pt]
= \frac{b^2}{2}\Bigg\{ & -\frac{\partial C_K}{\partial x}\frac{\Gamma_{12K} + \Gamma_{21K}}{2} \\
& + \left(C_{2K} - \frac{b}{2}\frac{\partial C_K}{\partial x}\right)\frac{\Gamma_{12K} - \Gamma_{21K}}{b} \\
& + \left(-\frac{\partial C_K}{\partial x}\right)\frac{\Gamma_{23K} + \Gamma_{32K}}{2} \\
& + \left(C_{2K} + \frac{b}{2}\frac{\partial C_K}{\partial x}\right)\frac{\Gamma_{23K} - \Gamma_{32K}}{b} \Bigg\} \qquad 2.26
\end{aligned}$$

The jump frequencies are also related. Consider the total jump rate from plane 2 as $\Gamma_K{}^1$.

$$\Gamma_K{}^1 = \frac{\Gamma_{23K} + \Gamma_{21K}}{2} \qquad 2.27$$

This implies that one may take

$$\Gamma_{12K} = \Gamma_{23K} - b\frac{\partial \Gamma_K{}^1}{\partial x}$$

and

$$\Gamma_{32K} = \Gamma_{21K} + b\frac{\partial \Gamma_K^1}{\partial x} \qquad 2.28$$

In addition, it is convenient to designate

$$\delta\Gamma_K^1 = \Gamma_{23K} - \Gamma_{21K} \qquad 2.29$$

which reflects the influence force and diffusion coefficient gradient combined. Hence in Eq. 2.29, one expects Eq. 2.23 to be applicable. Substituting Eqs. 2.27–2.29 into Eq. 2.26, one finds

$$J_K = \frac{b_2}{2}\left\{\frac{-\partial C_K}{\partial x}\left(\Gamma_K^1 - \frac{b}{2}\frac{\partial \Gamma_K^1}{\partial x}\right)\right.$$
$$+\left(C_{2K} - \frac{b}{2}\frac{\partial C_K}{\partial x}\right)\left(\frac{\delta\Gamma_K^1}{b} - \frac{\partial \Gamma_K^1}{\partial x}\right)$$
$$+\left(-\frac{\partial C_K}{\partial x}\right)\left(\Gamma_K^1 + \frac{b}{2}\frac{\partial \Gamma_K^1}{\partial x}\right)$$
$$+\left.\left(C_{2K} + \frac{b}{2}\frac{\partial C_K}{\partial x}\right)\left(\frac{\delta\Gamma_K^1}{b} - \frac{\partial \Gamma_K^1}{\partial x}\right)\right\}$$
$$=\frac{b^2}{2}\left\{-2\Gamma_K^1\frac{\partial C_K}{\partial x} - 2C_K\frac{\partial \Gamma_K^1}{\partial x} + 2C_K\frac{\delta\Gamma_K^1}{b} + \cdots\right\} \qquad 2.30$$

The terms in braces in Eq. 2.30, which are omitted, are on the order of b and, therefore, negligible. Hence, to first order, one has for long-range forces

$$J_K = -b^2\frac{\partial C_K\Gamma_K^1}{\partial x} + C_K b\delta\Gamma_K^1$$
$$= -\frac{\partial C_K D_K^*}{\partial x} + C_K V_K^1$$
$$= -\frac{b^2}{2}\frac{\partial C_K\Gamma_{0K}}{\partial x} + C_K V_K^1 \qquad 2.31$$

here $D_K^* = \frac{b^2}{2}\Gamma_{0K}$, $\Gamma_{0K} = 2\Gamma_K^1$, and $V_K^1 = b\delta\Gamma_K^1$. Equation 2.31 may be directly compared with Eq. 1.9. In that comparison it is convenient to substitute Eq. 2.23 using, Eq. 2.14 to evaluate the diffusion coefficient gradient. When one combines Eqs. 2.31, 2.23, and 2.14, the diffusion coefficient gradient term leading to δ_K in Eq. 2.14 cancels, since it is found in both $\partial D_K^*/\partial x$ and in V_K^1. Consequently, one may rewrite Eq.

2.31, using $V_{FK} = V_K{}^1 + \partial D_K^*/\partial x$ as

$$J_K = -D_K^* \frac{\partial C_K}{\partial x} + C_K V_{FK} \qquad 1.6$$

which is Eq. 1.6. The terms in $V_K{}^1$, which include δ_K from Eq. 2.14, are not present in Eq. 1.6; the diffusion coefficient gradient does not enter into the flux.

KINETIC THEORY FOR A DILUTE TERNARY fcc ALLOY

Suppose one considers two adjacent (100) planes in a three-dimensional fcc crystal oriented so that the planes will be perpendicular to an applied field, either a temperature gradient or electric potential gradient. The dilute substitutional solute is assumed to be immobile and vacancies are neglected. A dilute interstitial solute is placed on the octahedral interstitial sites. The X diffusion direction is also parallel with the applied field; concentration gradients of both the interstitial and substitutional solutes appear in the X direction. An alternate form for the flux compared to the earlier section is preferred in this section; the two forms are equivalent in the end.

One may denote these two planes as 1 and 2 and refer to the number of atoms per unit area of interstitial component on these planes as N_i, $i = 1, 2$. The planes are separated by a distance b, which is the X projection of the interstitial jump distance and one-half the dimension of the unit cell. Furthermore, one stipulates that Γ_{ij}, $i, j = 1, 2$, is the jump rate of interstitial component from plane i to plane j.

As a consequence of these definitions, the flux of the interstitial component is found to be

$$J = N_1\Gamma_{12} - N_2\Gamma_{21}$$

$$= (N_1 - N_2) \frac{\Gamma_{12} + \Gamma_{21}}{2}$$

$$+ \frac{N_1 + N_2}{2}(\Gamma_{12} - \Gamma_{21}) \qquad 2.32$$

With the interstitial concentration on plane i written as $C_i = N_i/b$ and C^1 as the average of the two, Eq. 2.32 may be written as

$$J = b^2\Gamma \frac{C_1 - C_2}{b} + bC^1(\Gamma_{12} - \Gamma_{21}) \qquad 2.33$$

where $\Gamma = (\Gamma_{12} + \Gamma_{21})/2$.

It is convenient to divide Eq. 2.33 by the number of interstitial sites per unit volume, M'. Then with $C^1/M' = C_{0'}$

$$\frac{J}{M'} = -b^2\Gamma\frac{\partial C_0}{\partial x} + bC_0(\Gamma_{12} - \Gamma_{21}) \qquad 2.34$$

C_0 is, therefore, the local fraction of occupied interstitial sites. This would imply that

$$C_1 = M'\left(C_0 - \frac{b}{2}\frac{\partial C_0}{\partial x}\right)$$

and $\qquad\qquad\qquad\qquad\qquad\qquad\qquad\qquad\qquad$ 2.35

$$C_2 = M'\left(C_0 + \frac{b}{2}\frac{\partial C_0}{\partial x}\right)$$

where C_0 may be associated with a pseudo-concentration halfway between planes 1 and 2. At this same position, I_0 could reflect the fraction of substitutional sites averaged between planes 1 and 2, which are occupied by the dilute substitutional component. The fraction of substitutional solute on plane at $\pm(b/2 + nb)$ for $n = 0, 1, 2, \cdots$ would be on the average

$$I_0\left[\pm\left(\frac{b}{2} + nb\right)\right] = I_0 \pm \left(\frac{b}{2} + nb\right)\frac{\partial I_{0'}}{\partial x} \qquad 2.36a$$

and

$$C_0\left[\pm\left(\frac{b}{2} + nb\right)\right] = C_0 \pm \left(\frac{b}{2} + nb\right)\frac{\partial C_0}{\partial x} \qquad 2.36b$$

Both the substitutional and interstitial solute concentrations are viewed as being sufficiently small that small complexes (pairs, triplets) are nearly negligible. Thus when the interstitial solute migrates it will only rarely combine with a second interstitial to form and interstitial pair. Define the fraction of interstitial sites occupied by such pairs as C_p. Furthermore, the migrating interstitial ion will with comparable frequency form an associated complex with an isolated substitutional solute. Thus I_p will be defined as the fraction of sites on either sublattice that contain an ion associated with such a pair.

The interstitial jump frequency will be influenced by those ions that are forming, moving, or dissociating the above pairs. A set of interstitial jump frequencies will consequently be defined to reflect such conditions. These definitions are modeled after the vacancy jump frequencies for a dilute binary substitutional solute to be discussed later. Let W_0 be the jump frequency of an interstitial ion that is isolated from other interstitials and from the substitutional solutes. The term W_1 is the jump rate where the interstitial remains a nearest neighbor to the substitutional; W_3 is the

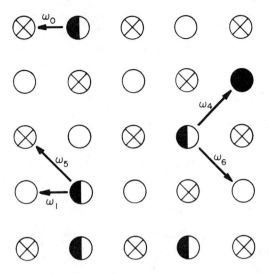

Figure 2.4 (100) Plane projection of interstitial jump sites and frequencies ($\bullet$ = substitutional solute; $\mathbb{O}$ = interstitial solute, $\otimes$ and $\bigcirc$ = solvent and interstitial sites on adjacent planes).

jump rate for dissociation of the interstitial-substitutional pair; and W_4 is the rate of association of such a pair. The term W_5 is the jump rate where an associated pair of interstitials dissociate, W_6 is the jump rate at which a pair of interstitials become associated, and W_7 is the jump frequency for bound interstitial pair migration. (See Fig. 2.4.)

When one defines C^* as the fraction of interstitial sites that contain isolated interstitial atoms, one can express the above in terms of two reactions. Use I^* for isolated substitutionals and these reactions are

$$C^* + C^* \underset{L2}{\overset{L1}{\rightleftharpoons}} C_p$$

$$C^* + I^* \underset{L4}{\overset{L3}{\rightleftharpoons}} I_p$$

2.37

These two equations lead to the following at equilibrium:

$$\frac{dI_p}{dt} = L_3 C^* I^* - L_4 I_p = 0$$

$$\frac{dC_p}{dt} = L_1 (C^*)^2 - L_2 C_p = 0$$

2.38

That these equations equal zero constitutes microscopic equilibrium. In Eq. 2.38, $L_1 = (7) \ (12) \ W_6$; $L_2 = (7) \ (2) \ W_5$; $L_3 = (6) \ (8) \ W_4$; and $L_4 = 8 W_3$. These rates imply that equilibrium is described by

$$C_p = \frac{6 W_6}{W_5} (C^*)^2,$$

and

$$I_p = \frac{6 W_4}{W_3} C^* I^* \qquad \qquad 2.39$$

When the conservation of mass is introduced,

$$C_0 = C^* + I_p + 2 C_p$$
$$I_0 = I^* + I_p \qquad \qquad 2.40$$

and one may solve for C_p and I_p. Since $I_p \ll I_0$ and $C_p \ll C_0$, one may approximate the results as follows:

$$Z_p = \frac{C_p}{12 C_0} \cong \frac{W_6}{2 W_5} C_0$$

$$Y_p = \frac{I_p}{6 C_0} \cong \frac{W_4}{W_3} I_0 \qquad \qquad 2.41$$

There are six distinguishable substitutional sites that one could find neighboring a given interstitial to form a pair belonging to I_p. Thus at equilibrium one defines Y_p as the probability that a particular one of them is filled. Similarly, given a known interstitial atom, there are 12 nearest neighbor positions on which an associated interstitial could reside. The term Z_p is therefore the probability that a given interstitial has a neighboring interstitial at one of these particular sites.

One may assume that such equilibrium exists even when a concentration gradient is present; the lack of such equilibrium violates microscopic equilibrium. With an applied field the probability terms Z_p and Y_p would be biased by the orientation of the complex with respect to the field; this bias is neglected herein. The neglect of such a contribution is justified. The pair association method of Lidiard suggests that these terms would lead to second-order corrections to the flux.

To calculate Γ_{12} for Eq. 2.33 and subsequently Γ_{21}, one views a diffusing interstitial on plane 1. The jump rate of this particular solute is dependent on the local solute concentration as well as the applied field. The local environment is to be represented by the probability that this interstitial has first or second neighbor solutes of the interstitial or substitutional kind. For example, with no long range applied field, there are four second neighbor substitutional solute sites a distance $3b/2$ from the position $x = 0$, and the diffusing interstitial is located at $-b/2$. The probability that any particular one of these sites is filled is found to be $I_0(3b/2)$. If any of these four sites has a substitutional solute, the interstitial solute jump towards such a solute would be W_4, and the jump rate to the other three sites is W_0. Thus there is a term in Γ_{12} of $4I_0(3b/2)$ $W_4 + 3W_0)$. Furthermore, the applied field biases the jump frequency to be $W_4^\pm = W_4(1 \pm \epsilon_4)$; the $\pm$ is due to the jump being parallel or antiparallel to the field. With the field parallel to the $+x$ direction, the positive sign would be applicable, and this modifies the previous contribution to become $4I_0(3b/2)[W_4(1+\epsilon_4)+3W_0(1+\epsilon_0)] = 4I_0(3b/2)[W_4^+ + 3W_0^+]$. Considering all first and second neighbor sites, one finds the jump frequency

$$\Gamma_{12} = 4I_0\left(\frac{3b}{2}\right)(W_4^+ + 3W_0^+) + 4I_0\left(\frac{b}{2}\right)(3W_4^+ + 5W_0^+)$$

$$+ Y_P\left(\frac{b}{2}\right)(4W_1^+) + 4Y_P\left(-\frac{b}{2}\right)(W_1^+ + 3W_3^+) + Y_P\left(-\frac{3b}{2}\right)(4W_3^+)$$

$$+ C_0\left(\frac{3b}{2}\right)(4W_6^+) + 4C_0\left(\frac{3b}{2}\right)(2W_6^+ + 2W_0^+)$$

$$+ 4C_0\left(\frac{3b}{2}\right)(W_6^+ + 3W_0^+) + 8C_0\left(\frac{b}{2}\right)(W_6^+ + 3W_0^+)$$

$$+ 4Z_P\left(\frac{b}{2}\right)(2W_7^+ + 2W_5^+) + 4Z_P\left(-\frac{b}{2}\right)(2W_7^+ + 2W_5^+)$$

$$+ 4Z_P\left(-\frac{3b}{2}\right)(4W_5^+) + 8I_0\left(-\frac{b}{2}\right)(4W_0^+) + 4I_0\left(-\frac{3b}{2}\right)(4W_0^+)$$

$$+ 4I_0\left(-\frac{5b}{2}\right)(4W_0^+) + 8C_0\left(-\frac{b}{2}\right)(4W_0^+) + 8C_0\left(-\frac{3b}{2}\right)(4W_0^+)$$

$$+ 4C_0\left(-\frac{5b}{2}\right)(4W_0^+) + C_0\left(-\frac{5b}{2}\right)(4W_0^+) + 4C_0\left(-\frac{5b}{2}\right)(4W_0^+)$$

$$+ 4W_0^+\left\{1 - 4I_0\left(\frac{3b}{2}\right) - 8I_0\left(\frac{b}{2}\right) - Y_P\left(\frac{b}{2}\right) - 4Y_P\left(-\frac{b}{2}\right)\right.$$

$$- Y_P\left(-\frac{3b}{2}\right) - 9C_0\left(\frac{3b}{2}\right) - 8C_0\left(\frac{b}{2}\right) - 4Z_P\left(\frac{b}{2}\right) - 4Z_P\left(-\frac{b}{2}\right)$$

$$- 4Z_P\left(-\frac{3b}{2}\right) - 8I_0\left(-\frac{b}{2}\right) - 4I_0\left(-\frac{3b}{2}\right) - 4I_0\left(-\frac{5b}{2}\right)$$

$$\left. - 8C_0\left(-\frac{b}{2}\right) - 8C_0\left(-\frac{b}{2}\right) - 8C_0\left(-\frac{3b}{2}\right) - 9C_0\left(-\frac{5b}{2}\right)\right\} \qquad 2.42$$

The first terms represent the product of the probability that the first or second neighbor solute is present times the jump rates if it were present. The last term in braces represents the jump rate if first or second neighbors were not present times the probability that no solute exists on first or second neighbor sites.

Equation 2.42 may be significantly simplified with some algebra and many terms cancel. Hence

$$
\Gamma_{12} = 4W_0{}^+ + 4I_0\left(\frac{3b}{2}\right)(W_4{}^+ - W_0{}^+) + 12I_0\left(\frac{b}{2}\right)(W_4{}^+ - W_0{}^+)
$$

$$
+ Y_p\left(\frac{b}{2}\right)(4W_1{}^+ - 4W_0{}^+) + 4Y_p\left(-\frac{b}{2}\right)(W_1{}^+ + 3W_3{}^+ - 4W_0{}^+)
$$

$$
+ Y_p\left(-\frac{3b}{2}\right)(4W_3{}^+ - 4W_0{}^+) + C_0\left(\frac{3b}{2}\right)(16W_6{}^+ - 16W_0{}^+)
$$

$$
+ 8C_0\left(\frac{b}{2}\right)(W_6{}^+ - W_0{}^+) + 4Z_p\left(\frac{b}{2}\right)(2W_7{}^+ + W_5{}^+ - 4W_0{}^+)
$$

$$
+ 4Z_p\left(-\frac{b}{2}\right)(2W_7{}^+ + 2W_5{}^+ - 4W_0{}^+) + 4Z_p\left(-\frac{3b}{2}\right)(4W_5{}^+ - 4W_0{}^+)
$$

2.43

In a similar manner one may find Γ_{21} as

$$
\Gamma_{21} = 4W_0{}^- + 4I_0\left(-\frac{3b}{2}\right)(W_4{}^- - W_0{}^-) + 12I_0\left(-\frac{b}{2}\right)(W_4{}^- - W_0{}^-)
$$

$$
+ 4Y_p\left(-\frac{b}{2}\right)(W_1{}^- - W_0{}^-) + 4Y_p\left(\frac{b}{2}\right)(W_1{}^- + 3W_3{}^- - 4W_0{}^-)
$$

$$
+ 4Y_p\left(\frac{3b}{2}\right)(W_3{}^- - W_0{}^-) + 16C_0\left(-\frac{3b}{2}\right)(W_6{}^- - W_0{}^-)
$$

$$
+ 8C_0\left(-\frac{b}{2}\right)(W_6{}^- - W_0{}^-) + 4Z_p\left(-\frac{b}{2}\right)(2W_7{}^- + W_5{}^- - 4W_0{}^-)
$$

$$
+ 8Z_p\left(\frac{b}{2}\right)(W_7{}^- + W_5{}^- - 2W_0{}^-) + 16Z_p\left(\frac{3b}{2}\right)(W_5{}^- - W_0{}^-) \qquad 2.44
$$

Further algebra is facilitated by the combination of Eqs. 2.36 and 2.41; thereby

$$
Z_p\left(\pm\left(\frac{b}{2} + nb\right)\right) = \frac{W_6}{2W_5}\left\{C_0\left(\pm\left(\frac{b}{2} + nb\right)\right)\right\}
$$

$$
Y_p\left(\pm\left(\frac{b}{2} + nb\right)\right) = \frac{W_4}{W_3}\left\{I_0\left(\pm\left(\frac{b}{2} + nb\right)\right)\right\}
$$

2.45

Upon combining Eqs. 2.36, 2.43, 2.44, and 2.45, one finds the average frequency

$$\Gamma = \frac{\Gamma_{12} + \Gamma_{21}}{2} = 4\left\{ W_0 + I_0\left[4(W_4 - W_0) \right.\right.$$

$$\left. + 2\frac{W_4}{W_3}(W_1 + 2W_3 - 3W_0) \right]$$

$$\left. + C_0\left[6(W_6 - W_0) + \frac{W_6}{2W_5}(4W_7 + 7W_5 - 12W_0) \right]\right\} \qquad 2.46$$

Furthermore, the frequency difference

$$\Gamma_{12} - \Gamma_{21} = 8b\left\{ \frac{\partial I_0}{\partial x}\left[3(W_4 - W_0) + \frac{3W_4}{W_3}(W_0 - W_3) \right]\right.$$

$$+ \frac{\partial C_0}{\partial x}\left[7(W_6 - W_0) + \frac{W_6}{W_5}\left(3W_0 - \frac{13W_5}{4} \right) \right] + W_0\epsilon_0$$

$$+ 2\frac{W_4}{W_3} I_0(2W_1\epsilon_1 + 4W_3\epsilon_3 - 6W_0\epsilon_0) + 4I_0(W_4\epsilon_4 - W_0\epsilon_0)$$

$$\left. + 6C_0(W_6\epsilon_6 - W_0\epsilon_0) + \frac{W_6C_0}{2W_5}(4W_7\epsilon_7 + 7W_5\epsilon_5 - 12W_0\epsilon_0) \right\} \qquad 2.47$$

In Eq. 2.47, the ϵ_i depends on the type of applied field, a temperature gradient or an electric potential gradient. Thus for an electric field one has $\epsilon_i = Eq/2K_BT$, where E is the field intensity and q is the effective charge on the interstitial ion. For a temperature gradient, $\epsilon_i = (-q_i^*/2K_BT^2)\nabla T$ with q_i^* the heat of transport for a jump at rate W_i.

The activity coefficient gradients are not introduced into Eq. 2.47. The activity reflects the immediate atomic neighborhood of the diffusing ion, whereas the externally applied field terms are macroscopic in nature. One may understand this difference as follows: For an infinitely dilute solution the activity coefficient reflects the constant enthalpy and entropy of mixing isolated atoms (Henry's law). As the concentration rises, pair formation becomes probable and a departure from Henry's law is found. This gives rise to an influence on the flux as seen below. At still higher concentrations an analysis of these terms becomes very complicated for a kinetic theory of higher order complexes. With the jump frequencies previously defined, this analysis is roughly comparable with a first neighbor quasi-chemical theory of pair formation.

With Eqs. 2.34, 2.46, and 2.47, one may determine an expression for the flux of interstitial solute as follows:

$$-\frac{J}{M'} = b^2\Gamma\left[\frac{\partial C_0}{\partial x} + \frac{C_0 A}{\Gamma}\frac{\partial C_0}{\partial x} + \frac{C_0 B}{\Gamma}\frac{\partial I_0}{\partial x}\right.$$

$$\left. + \frac{C_0 Q}{K_B T^2}\nabla T - \frac{C_0 Q_e E}{K_B T}\right] \qquad 2.48a$$

where

$$A = -8\left[7(W_6 - W_0) + \frac{W_6}{W_5}\left(3W_0 - \frac{13W_5}{4}\right)\right]$$

$$B = -8\left[3(W_4 - W_0) + \frac{3W_4}{W_3}(W_0 - W_3)\right]$$

$$Q = \frac{8}{\Gamma}\left\{W_0 q_0^* + 4I_0(W_4 q_4^* - W_0 q_0^*) + \frac{W_4 I_0}{W_3}(2W_1 q_1^*\right.$$

$$+ 4W_3 q_3^* - 6W_0 q_0^*) + 6C_0(W_6 q_6^* - W_0 q_0^*)$$

$$\left. + \frac{W_6 C_0}{2W_5}(4W_7 q_7^* + 7W_5 q_5^* - 12W_0 q_0^*)\right\}$$

and Q_e is the same expression as Q, except that $q = q_i^*$ for all i.

An alternate form of Eq. 2.48a is preferable in what follows: Let

$$4\Gamma_0 = \Gamma + C_0\left\{\frac{6W_6}{W_5}W_0 + 2W_0 - \tfrac{15}{2}W_6\right\}$$

and

$$4A_0 = A - \left\{\frac{6W_6 W_0}{W_5} + 2W_0 - \tfrac{15}{2}W_6\right\}$$

so that when E and ∇T are zero, one may write Eq. 2.48a as follows:

$$-\frac{J}{M'} = 4b^2\Gamma_0\left\{\frac{\partial C_0}{\partial x} + \frac{C_0 A_0}{\Gamma_0}\frac{\partial C_0}{\partial x} + \frac{C_0 B}{\Gamma_0}\frac{\partial I_0}{\partial x}\right\} \qquad 2.48b$$

Several distinctive aspects arise in Eq. 2.48a. First, one notices that the heat of transport for thermal diffusion, Q, and the effective charge of electromigration are concentration-dependent; the origin for this concentration dependence is pair formation of either the interstitial or substitutional-interstitial type.

Second, and perhaps more importantly is the existence of the second two terms in Eq. 2.48. The origin of these terms is also pair formation.

These terms, A and B, give rise to a velocity of the interstitial solute based on the different jump frequencies associated with the formation or dissociation of pairs. These terms play the same role as the activity coefficient gradient in a phenomenological theory. It is not unreasonable, therefore, to attempt to construct a phenomenological theory and identify the nature of the comparison. We shall, therefore, construct the phenomenological theory based on pair formation and dissociation as an influence on the diffusive flux. The flux is assumed to follow the equation $-(J/M') = \sum_K L_{iK}(\partial \mu_K/\partial x)$, where μ_K is the chemical potential of component K.

Chemical Potential

To determine the chemical potential one may assume that the off-diagonal phenomenological coefficients, the L_{ij} for $i \neq j$ terms, are zero in this model because the substitutional solute is immobile and the Onsager reciprocal relations should hold. The last section of this chapter discusses the situation where L_{ij} are nonzero. Therefore, when $E = \nabla T = 0$, the flux is assumed to be given by

$$-\frac{J}{M'} = L_{11}\frac{\partial \mu_1}{\partial x} = L_{11}\frac{\partial \mu}{\partial C_0}\frac{\partial C_0}{\partial x} + L_{11}\frac{\partial \mu_1}{\partial I_0}\frac{\partial I_0}{\partial x}$$

$$= \frac{L_{11}K_BT}{C_0}\frac{\partial C_0}{\partial x} + L_{11}KT\frac{\partial \ln \gamma}{\partial C_0}\frac{\partial C_0}{\partial x}$$

$$+ L_{11}K_BT\frac{\partial \ln \gamma}{\partial I_0}\frac{\partial I_0}{\partial x} \qquad 2.49$$

The first term in both Eqs. 2.48 and 2.49 are associated with the entropy of mixing of an ideal solution and, therefore, must be equal. Consequently, by comparison

$$L_{11} = \frac{4b^2\Gamma_0 C_0}{K_B T} \qquad 2.50$$

The second two terms in both Eqs. 2.48 and 2.49 are linearly independent. Consequently,

$$\frac{A_0}{\Gamma_0} = \frac{\partial \ln \gamma}{\partial C_0}$$

and

$$\frac{B}{\Gamma_0} = \frac{\partial \ln \gamma}{\partial I_0} \qquad 2.51$$

Equations 2.51 constitute a pair of simultaneous, partial differential equations. Since Γ_0 is of the form $W_0 + EC_0 + FI_0$ where E and F are functions of W_i, Eq. 2.51 may be simplified by the assumption that E and F are nearly zero. Thus we find from Eq. 2.46,

$$F = 3(W_4 - W_0) + 2\frac{W_4}{W_3}(W_1 + 2W_3 - 3W_0) \cong 0$$

$$E = -\tfrac{3}{2}W_6 - 4W_0 + \frac{W_6}{2W_5}(4W_7 + 7W_5) \cong 0 \qquad\qquad 2.52$$

When Eq. 2.52 is combined with Eq. 2.51,

$$\frac{\partial \ln \gamma}{\partial C_0} \cong \frac{A_0}{W_0}\left\{1 - \frac{E}{W_0}C_0 - \frac{F}{W_0}I_0\right\} \cong \frac{A_0}{W_0}$$

$$\frac{\partial \ln \gamma}{\partial I_0} \cong \frac{B}{W_0}\left\{1 - \frac{EC_0}{W_0} - \frac{F}{W_0}I_0\right\} \cong \frac{B}{W_0} \qquad\qquad 2.53$$

because C_0 and I_0 are small in addition to E and F. The final result for Eq. 2.53 can be integrated to the form

$$\ln \gamma = \ln \gamma_0 + \frac{A_0}{W_0}C_0 + \frac{B}{W_0}I_0 \qquad\qquad 2.54$$

where γ_0 is the activity coefficient at infinite dilution (Henry's law). Equation 2.54 appears as an unusual form for the activity coefficient. In the following section it is shown from statistical mechanics that Eq. 2.54 is a very reasonable equation for the activity coefficient. First, however, the chemical potential is written from Eq. 2.54. Evidently, this theory gives

$$\mu = K_B T\left\{\ln C_0\gamma_0 + C_0\left(12 - 12\frac{W_6}{W_5}\right) + I_0\left(6 - \frac{6W_4}{W_3}\right)\right\} \qquad\qquad 2.55$$

Statistical Mechanics of a Dilute Ternary Alloy

To evaluate the chemical potential of the dilute ternary component, the canonical ensemble partition function for the previously described system must be evaluated with approximations similar to those that yield Eq. 2.55. To this end, the following definitions are important: N_1 is the number of interstitial atoms and M is the number of solvent atoms. Assume throughout this section that $N \gg N_1$, $N \gg M$, and $N + M \cong N$ lattice sites. Furthermore, the following definitions define the appropriate bond energies. Let E_0 be the energy of a solvent–solvent bond; E_1 be the energy of an isolated interstitial in the solvent lattice; E_2 be energy of a substitutional solute–solvent bond; E_3 be the energy of an interstitial–interstitial pair; and E_4 be the energy of an interstitial–substitutional solute pair.

Excluding the interstitials, the lattice energy is E_L and may be found from

$$E_L = 6(N - M)E_0 + 12ME_2 \qquad 2.56$$

One further defines K_1 as the number of isolated interstitials, K_2 as the number of interstitial–interstitial pairs, and K_3 as the number of interstitial–substitutional solute pairs. With these definitions the energy of the sublattice on which the interstitials are found as E_{SL}.

$$E_{SL} = K_1 E_1 + K_2 E_3 + K_3 E_4 \qquad 2.57$$

The total energy within the system is $E_L + E_{SL}$.

The degeneracy of the system may be divided into that portion associated with the lattice, $\Omega_L(N, M)$, and that portion associated with the sublattice, $\Omega_{SL}(N, M, N_1, K_1, K_2, K_3)$. The total degeneracy Ω is then found from

$$\Omega = \Omega_L \Omega_{SL} \qquad 2.58$$

Now since the lattice energy and degeneracy are only dependent on N and M, the partition function of the lattice will not influence the free energy of the sublattice. That is, all interaction effects will be associated with the sublattice on which the interstitials are found. Thus the chemical potential of the interstitial may be determined from the partition function for the sublattice, Q_S, as follows:

$$Q_S = \sum_{K_1, K_2, K_3} \Omega_{SL} \exp -\frac{\{K_1 E_1 + K_2 E_3 + K_3 E_4\}}{K_B T} \qquad 2.59$$

where the sum has the restriction that $K_1 + 2K_2 + K_3 = N_1$. This restriction is placed later as necessary.

To evaluate the degeneracy is a complicated problem even in the approximation that only pairs are formed. To the author's knowledge an exact expression is unknown. A reasonable approximation is developed below as evidenced by the results it yields.

First, subdivide the interstitial solute atoms into three distinguishable groups numbering K_1, $2K_2$, and K_3, associated with the respective energies. With this subdivision, the sublattice will be divided into three parts; those sublattice sites that neighbor the substitutional solute atoms and those that do not. The sites that do not neighbor the substitutional solute atoms are subdivided into sites on which only isolated interstitials are placed on sites when pairs may be placed with no overlap of lattice sites.

By hypothesis, $M > K_3$, and there are six sites surrounding each substitutional solute where an interstitial may be placed. Define Ω_3 as the number of ways of placing the K_3 atoms on the six M sites so that only one interstitial pair may be formed from each substitutional solute. Then, one may calculate Ω_3 from

$$\Omega_3 = \frac{6M(6M-6)(6M-12)\cdots[6M-6(K_3-1)]}{K_3!}$$

$$= \frac{6^{K_3}M!}{K_3!\,(M-K_3)!} \tag{2.60}$$

There are $N + M - 6M$ remaining interstitial sites on which $K_1 + 2K_2$ atoms will be placed. The K_1 atoms are viewed as being isolated so that each of these atoms might be expected to exclude 13 sites; the one on which it is placed plus the 12 nearest neighbors. If, however, only 12 sites are excluded for each atom belonging to K_1, one may contemplate interchange of some of the $2K_2$ atoms forming pairs within the group containing K_1 atoms. The subdivision between these groups is necessary because the atoms are not distinguishable. Let Ω_1 be the number of ways of placing the K_1 atoms on their respective sites. To this end, let $B = (N - 5M)/12$, then Ω_1 is determined from

$$\Omega_1 = \frac{(12B)(12B-12)(12B-24)\cdots[12B-12(K_1-1)]}{K_1!}$$

$$= \frac{(12)^{K_1}B!}{K_1!\,(B-K_1)!} \tag{2.61}$$

There are $12(B - K_1)$ remaining interstitial sites on which interstitial pairs may be placed. Now each interstitial pair occupies 2 sites and there are at least 18 empty nearest neighbor sites to the pair. Hence to exclude the formation of triplets, one must exclude 20 sites for each pair so formed. Also, if the first member of the first pair may be placed on $12(B - K_1)$ sites, the second member of the pair is found on one of the 12 sites adjacent to the first. Hence there are Ω_2 ways to place the K_2 pairs on these sites. Let $20D = 12(B - K_1)$, so that

$$\Omega_2 = \frac{(20D)(12)(20D-20)(12)\cdots[20D-(K_2-1)20](12)}{2^{K_2}K_2!}$$

$$= \frac{[(20)(6)]^{K_2}D!}{K_2!\,(D-K_2)!} \tag{2.62}$$

When one evaluates Eq. 2.71 in the limit where $N \gg N_1$, which also implies $N \gg K_2^*$ and $N \gg K_3^*$, one finds

$$\frac{\mu}{K_B T} = \ln \frac{N_1}{N+M} + \frac{E_1}{K_B T} + \frac{N_1}{N+M} (12 - 12 e^{H_2/K_B T})$$

$$+ \frac{M}{N+M} (6 - 6 e^{H_3/K_B T}) \qquad\qquad 2.72$$

Approximations of the nature

$$\ln \left(1 + \frac{N_1}{N} + \frac{M}{N} + \frac{2K_2}{N} + \frac{K_3}{N} \right) = \frac{N_1}{N} + \frac{M}{N} + \frac{2K_2}{N} + \frac{K_3}{N}$$

are used in Eq. 2.72. One must also assume that $K_2 \ll N_1$, that $K_3 \ll N_1$, and that only linear terms in N_1 and M are retained. Using the previously defined jump frequencies, and substituting $\frac{M}{N+M} = I_0$, $\frac{N_1}{N+M} = C_0$, one finds Eq. 2.41 to be

$$\frac{\mu}{K_B T} = \ln \gamma_0 C_0 + C_0 \left(12 - 12 \frac{W_6}{W_5} \right) + I_0 \left(6 - 6 \frac{w_4}{W_3} \right) \qquad\qquad 2.73$$

with $\ln \gamma_0 = E_1/K_B T$. Equations 2.73 and 2.55 are identical.

The tendency for interstitial-substitutional pair formation occurs when $W_4/W_3 \gg 1$. Thus B is clearly negative in Eq. 2.48 whenever $W_0 \cong W_3$, $W_4 \gg W_3$, $W_4 \cong W_0$, and $W_3 \ll W_4$. If, for example, $\partial C_0/\partial x \cong 0$, this condition could lead to a flux of the interstitial component toward the region that is rich in the substitutional solute. This "uphill" diffusion, $D_{12} < 0$, is a manifestation of the chemical potential of the interstitial solute to be decreased by segregation to the region that is rich in the substitutional solute. The same conclusion may be reached through Eq. 2.55. Similarly, Eqs. 2.48 and 2.55 would yield the flux away from a region rich in the substitutional solute when $W_4/W_3 \ll 1$ for the conditions $W_0 \cong W_3$ and $W_4 \ll W_3$, or $W_4 > W_0$ and $W_3 \gg W_4$. It is also notable that B is zero when $W_0 = W_3$ and W_4, so that the substitutional solute does not influence the interstitial in this instance.

It is obvious that similar comments may be made about the term A_0 in Eqs. 2.48 and 2.55 for $W_6/W_5 \gg 1$ or $W_6/W_5 \ll 1$. There would be an increasing tendency for "uphill" interstitial solute diffusion when solute pairs are stable. When $W_6/W_5 \gg 1$, the chemical potential decreases with interstitial segregation.

This kinetic model of diffusion of an interstitial solute in a dilute ternary alloy has been developed for the case where interstitial pair and interstitial-substitutional pair formation influence the flux. These terms give rise to a chemical potential gradient that can cause "uphill" diffusion. The activity coefficient gradient is a manifestation of the pair and higher order complex formation and not a long range force such as exists with a temperature or electric field gradient. In the presence of these latter gradients, the tendency for pair formation leads to a concentration dependence for the effective charge or heat of transport.

PHENOMENOLOGICAL EQUATIONS

In the previous two sections it was affirmed that the driving force for diffusion of a dilute interstitial was its chemical potential gradient. This effect was a simplification of the phenomenon of diffusion in a general alloy containing an interstitial solute because the substitutional impurity was considered immobile. The general situation where all components can move presents a complication that can conveniently be dealt with in a phenomenological manner. Interestingly, the equation derived for the flux of the dilute interstitial is very close to the result if the substitutional impurity were mobile. The flux equation, 2.48b, for the situation in which ∇T and E were zero, may be written as

$$J_1 = -D_{11}\frac{\partial C_1}{\partial x} - D_{12}\frac{\partial C_2}{\partial x} \qquad 2.74$$

where component 1 is the interstitial and component 2 is the substitutional solute. The form for this equation is correct although the coefficients D_{ij} would not be exactly those given in Eq. 2.48b. One may write another equation for the substitutional component.

$$J_2 = -D_{22}\frac{\partial C_2}{\partial x} - D_{21}\frac{\partial C_1}{\partial x} \qquad 2.75$$

Equations 2.74 and 2.75 are correct for the mass transport when only concentration varies with position. The diffusivities D_{ij} are functions of concentration and are not necessarily positive numbers. Since the diffusion coefficients are not necessarily positive, one can have diffusion up concentration gradients, which leads to discontinuities in concentration. This is a disorderly situation to consider mathematically, as in Chapter 1. One may avoid such complications by considering an alternate form for Eqs. 2.74 and 2.75 in which chemical potential gradients are present. That this is the proper choice for the driving force is emphasized by the results of the previous two sections.

Since the chemical potentials may be taken as independent thermodynamic variables as a substitute for the concentrations as indicated in Eqs. 2.74 and 2.75. Thus

$$\frac{\partial C_j}{\partial x} = \sum_{K=1}^{2} \frac{\partial C_j}{\partial \mu_K} \frac{\partial \mu_K}{\partial x} \qquad 2.76$$

Substituting Eq. 2.76 into either Eq. 2.74 or 2.75

$$J_i = -\sum_{j=1}^{2} \sum_{K=1}^{2} D_{ij} \frac{\partial C_j}{\partial \mu_K} \frac{\partial \mu_K}{\partial x} = -\sum_{K=1}^{2} L_{iK} \frac{\partial \mu_K}{\partial x} \qquad 2.77$$

where i equals 1 or 2 and

$$L_{iK} = \sum_{j=1}^{2} D_{ij} \frac{\partial C_j}{\partial \mu_K} \qquad 2.78$$

The L_{iK} are called phenomenological coefficients. In the kinetic theory described in the section before last, we found

$$L_{11} = \dot{D}_{11} \frac{\partial C_1}{\partial \mu_1} + D_{12} \frac{\partial C_2}{\partial \mu_2} \qquad 2.79a$$

$$L_{12} = D_{11} \frac{\partial C_1}{\partial \mu_2} + D_{12} \frac{\partial C_2}{\partial \mu_2} \qquad 2.79b$$

The L_{12} was zero is a consequence of the particular model as follows:

$$J_2 = -L_{21} \frac{\partial \mu_1}{\partial x} - L_{22} \frac{\partial \mu_2}{\partial x} \qquad 2.80$$

was zero by hypothesis and the substitutional component was immobile. Since $\partial \mu_2/\partial x$ and $\partial \mu_1/\partial x$ are arbitrary, $L_{22} = L_{21} = 0$. As will be seen, $L_{21} = L_{12}$; hence $L_{12} = 0$.

It is the purpose of this section to generalize the Eq. 2.77 to a situation where one may consider arbitrary forces of which the chemical potential gradients are a proper choice. The advantage of the chemical potential gradient as the proper choice for the forces lies in the laws of thermodynamics. It is known from thermodynamics that the chemical potential of a component in a particular phase is the same number for some other phase that is in equilibrium with the first. Thus the chemical potential of a component in an arbitrary alloy will be constant at equilibrium for no temperature or pressure gradients. It is not surprising that when the system is out of equilibrium, the chemical potentials of the components quickly become continuous functions. The concentrations are not continuous across phase boundaries. Consequently, it is of great practical interest to re-express the fluxes in terms of chemical potential gradients instead of concentration gradients; the former will become continuous in the early stages of diffusion. In the process of writing the fluxes in terms of chemical potential gradients, one arrives at the generalized form for the flux equations for any irreversible process of which diffusion mass fluxes are just one example.

To treat this problem, one must first consider the thermodynamics of an irreversible process in which diffusive fluxes exist. The major quantity of interest is the entropy production rate per unit volume, σ. This is a positive definite and tends to zero as equilibrium is approached.

In an arbitrary volume of the system, V, the total entropy S is changed by two processes, entropy production and entropy flow. The entropy flow is associated among other things with the flow of matter and its intrinsic entropy content. The rate of entropy change for the volume V, with ρ the mass density, is therefore

$$\frac{dS}{dt} = \int_V \frac{\partial \rho \underline{S}}{\partial t} \, dV \qquad 2.81$$

$\underline{S}$ is the entropy per unit mass inside the system. The entropy that flows through the surface Ω of the volume V is contained in an entropy flux term, $J_{S,t}$. Hence the rate of change of entropy

$$\frac{dS}{dt} = \int_V \frac{\partial \rho \underline{S}}{\partial t} = -\int_\Omega J_{S,t} \, d\Omega + \int_V \sigma \, dV \qquad 2.82$$

The total entropy flux $J_{S,t}$ is related to the entropy flow J_S through

$$J_S = J_{S,t} - \rho \underline{S} v_m \qquad 2.83$$

where v_m is the velocity of the local center of mass. On combining Eqs. 2.82 and 2.83, the Gauss theorem, and the definition of the substantial time derivative where

$$\frac{d}{dt} = \frac{\partial}{\partial t} + v_m \cdot \nabla$$

Eq. 2.82 may then be written as

$$\frac{\rho \, d\underline{S}}{dt} = -\nabla J_S + \sigma \qquad 2.84$$

Equation 2.84 is a statement of the rate of change in the entropy for an arbitrary irreversible process. Those processes that contribute to σ and are of interest herein are principally the flows of charge, heat, and matter. There can be fluxes of any component K in the system J_K, a charge flux, J_e, and a heat flux, J_q. The forces causing these phenomena are the chemical potential gradient of component K, $-\nabla\mu_K/T$, the electric potential gradient, E, and the temperature gradient, $-\nabla T/T$. If one denotes the force x_i and the associated or conjugate flux J_i, one may express the entropy production rate per unit volume as

$$T\sigma = \sum_i J_i x_i \qquad\qquad 2.85$$

One may use Eq. 2.85 to develop equations of the form 2.77 in the following manner. The term $T\sigma$ is known to be a positive quantity that becomes zero as equilibrium is approached. When equilibrium is approached, the forces x_i, and the fluxes, must approach zero. Letting the subscript e denote equilibrium, this implies that

$$T\sigma\bigg|_e = 0 \qquad\qquad 2.86$$

Furthermore,

$$\frac{\partial T\sigma}{\partial x_j}\bigg|_e = J_j\bigg|_e + \sum_i x_i \frac{\partial J_i}{\partial x_j}\bigg|_e = 0 \qquad\qquad 2.87$$

since x_i and J_j are zero when no forces are present.

Based on Eqs. 2.86 and 2.87 it seems reasonable to take a MacLaurin expansion in $T\sigma$ about equilibrium using the variables x_i. Hence

$$T\sigma = T\sigma\bigg|_e + \sum_j \frac{\partial T\sigma}{\partial x_j}\bigg|_e x_j$$
$$+ \frac{1}{2}\sum_i\sum_j \frac{\partial^2 T\sigma}{\partial x_i \partial x_j}\bigg|_e x_i x_j + \cdots$$

$$2.88$$

Such expansion is valid in the absence of a magnetic field. Comparing Eq. 2.88 with Eqs. 2.86 and 2.87,

$$T\sigma = \frac{1}{2}\sum_i\sum_j \frac{\partial^2 T\sigma}{\partial x_i \partial x_j}\bigg|_e x_i x_j + \cdots \qquad\qquad 2.89$$

If one retains only the first nonzero term, then for irreversible processes near equilibrium one may deduce the flux terms in Eq. 2.85 by comparing it with Eq. 2.89. This leads one to

$$J_i = \frac{1}{2}\sum_j \frac{\partial^2 T\sigma}{\partial x_i \partial x_j}\bigg|_e x_j$$

$$= \sum_j L_{ij} x_j \qquad\qquad 2.90$$

Equations 2.90 and 2.77 have the exact same form since the chemical potential gradients $-\partial \mu_K/\partial x$ are written as x_K in Eq. 2.90. The terms L_{iK} in Eq. 2.90 are known as phenomenological coefficients. Evidently,

$$L_{iK} = \frac{1}{2} \frac{\partial^2 T\sigma}{\partial x_i \partial x_K}\bigg|_e \qquad 2.91$$

Thus in the absence of a magnetic field, one has

$$L_{iK} = \frac{1}{2} \frac{\partial^2 T\sigma}{\partial x_i \partial x_K}\bigg|_e = \frac{1}{2} \frac{\partial^2 T\sigma}{\partial x_K \partial x_i}\bigg|_e$$

$$= L_{Ki} \qquad 2.92$$

Equations 2.92 are known as the Onsager reciprocal relations. Such a relationship between the phenomenological coefficients has a profound effect on experiments. A heat flux is well known to arise from a temperature gradient. The nonzero phenomenological coefficients would assert that a mass flux would also be expected when a temperature gradient is present. The reciprocal relations would further stipulate that a chemical potential gradient, which ordinarily yields a mass flux, would also create a heat flow. In gases this phenomena is measurable; in solids it is too small for experimental determination.

Equations 2.92 lead one to expect a whole range of phenomena to occur when a system is out of equilibrium. For example, the diffusive fluxes of components in an alloy would follow

$$J_i = -L_{i1} \frac{\nabla \mu_1}{T} - L_{i2} \frac{\nabla \mu_2}{T} + \cdots$$

$$- L_{iq} \frac{\nabla T}{T} + \cdots, \qquad i = 1, 2, 3, \cdots N \qquad 2.93$$

for an N-component system.

The heat flux would also be written in a similar manner

$$J_q = -L_{q1} \frac{\nabla \mu_1}{T} - L_{q2} \frac{\nabla \mu_2}{T} + \cdots$$

$$- L_{qq} \frac{\nabla T}{T} \qquad 2.94$$

The amazing thing about Eq. 2.92 is that $L_{i1} = L_{1i}$, $L_{i2} = L_{2i}$, $\cdots$, and even more interesting, $L_{iq} = L_{qi}$. The chemical potential gradient causes the same heat flow as a temperature gradient causes a flow of matter.

The phenomenological coefficients and the reciprocal relations were introduced to generalize the flux equation to the response from the thermodynamic driving force. Any phenomenological theory has its merit in the generality that it presents. However, one is left without the details of the kinetic processes that may occur; these processes are responsible for the phenomenological coefficients being nonzero. Earlier in this chapter, the cross effects between an electric field and temperature gradient with the flux of an interstitial solute were identified. Also, interstitial pair and interstitial-substitutional pair formation were introduced. These interactions detailed the origin of the chemical potential gradient. Other complexes in the solid solution could be present in sufficient quantity to influence the flux through nonzero contributions to chemical potential gradients or phenomenological coefficients.

When the vacancies in the previous ternary alloy are considered, the substitutional solute and the solvent become mobile by a vacancy mechanism. This leads to a number of complexes that could form to influence the flux of the interstitial. The first possibility is a vacancy that wanders up to an isolated interstitial. The presence of the interstitial alters the vacancy jump frequency; or equally likely, the vacancy alters the interstitial jump frequency. Second, such a complex could lead to the annihilation of the vacancy by the interstitial assuming a substitutional site. Again the fluxes are altered by the formation of the complex. These two phenomena could also occur when one or more of the vacancies neighbors is a substitutional solute. The annihilation of a substitutional-vacancy pair by an interstitial solute jumping into the vacancy would certainly change the jump rate of the substitutional solute; hence the presence of the interstitial would decrease the flux of the substitutional solute by the annihilation of the vacancy into which the substitutional could jump. The complex so formed would also decrease the mobility of the interstitial since the normally mobile interstitial would be trapped momentarily in a substitutional site. Such trapping certainly leads to the presence of off-diagonal phenomenological coefficients, L_{iK}, $i \neq K$, since the presence and, therefore, chemical potential of one solute influences the mobility of another solute.

Such trapping phenomena are also expected in a ternary alloy with two interstitial solutes. The immediate presence of interstitial solute A would change the jump rates of a neighboring interstitial solute B. The changed jump rate gives a contribution to the chemical potential. Also, an A-B pair of interstitials would be competing for the same empty site during jumps. Hence A would hold up B jumps, and vice versa. This type of interaction certainly leads to the L_{iK}, $i \neq K$, being nonzero in the phenomenological equations.

If any of these mechanisms are sufficiently operative that they influence the flux, then it is important to include the flux of vacancies and the chemical potential of the vacancies as if these defects were another component. In this instance, Eq. 2.93 would have $N+1$ fluxes and $N+1$ components, the last component being vacancies. Letting $\nabla \mu_V|_T$ be the chemical potential gradient of vacancies, and J_V be the vacancy flux, one would substitute the following set of equations for Eqs. 2.93 and 2.94.

$$J_i = -L_{i1} \nabla \mu_1|_T - L_{i2} \nabla \mu_2|_T \cdots -L_{iN} \nabla \mu_N|_T$$

$$- L_{iV} \nabla \mu_V|_T - L_{iq} \frac{\nabla T}{T}, \; i = 1, \cdots N,$$

$$J_V = -L_{V1} \nabla \mu_1|_T - L_{V2} \nabla \mu_2|_T, \cdots -L_{VN} \nabla \mu_N|_T$$

$$-L_{VV} \nabla \mu_V|_T - L_{Vq} \frac{\nabla T}{T},$$

and

$$J_q = -L_{q1} \nabla \mu_1|_T - L_{q2} \nabla \mu_2|_T \cdots -L_{qN} \nabla \mu_N|_T$$

$$-L_{qV} \nabla \mu_V|_T - L_{qq} \frac{\nabla T}{T} \qquad\qquad 2.95$$

A set of equations such as Eq. 2-95 is mandatory if the solutes are substitutional and diffuse by a vacancy mechanism. For interstitial solutes the vacancy flux and chemical potential gradient would only be important when interstitial vacancy complexes are prevalent. The formulation of a kinetic theory to include higher-order complexes but along parallel lines with the one presented earlier in the chapter would be very difficult.

SOME EXPERIMENTAL RESULTS

If one takes the limit of Eq. 2.48a for the situation that $I_0 = 0$ and $C_0 \to 0$, one finds for $E = \nabla T = 0$ that

$$J = -4b^2 W_0 \frac{\partial M' C_0}{\partial x}$$

$$= -D^* \frac{\partial M' C_0}{\partial x} \qquad\qquad 2.96$$

Thus $D^* = 4b^2 W_0$. From Eq. 2.13 $W_0 = \nu \exp(-\Delta G/K_B T)$ and, therefore,

$$D^* = 4b^2 \nu \exp\left(\frac{\Delta S}{K_B} - \frac{\Delta H}{K_B T}\right) = D_0 \exp\left(\frac{-\Delta H}{K_B T}\right) \qquad\qquad 2.97$$

Thus our theory predicts that a plot of $\ln D$ versus $1/T$ will yield a straight line with slope $-\Delta H/K_B$. Experimentally, a straight line has been found on such a plot for such systems as carbon in iron or nitrogen in niobium over as many as five orders of magnitude in the diffusivity.

For diffusion of interstitial solutes with an applies electric field, Eq. 1.10 would imply that when $t \to \infty$ a steady state concentration would be found which satisfies

$$\frac{\partial \ln C}{\partial x} = \frac{Q_e E}{K_B T} \qquad 2.98$$

Thus a steady state concentration gradient can be obtained with an applied electric field from which Q_e may be obtained. A temperature gradient is usually present in such cases also, and interpretation of data is therefore more difficult. Nevertheless, the use of either Eq. 2.98 or Eq. 1.34 yields measured values for Q_e. For example, carbon in austenitic iron yielded a value of $Q_e = 3.7$ electron charges[1] nearly consistent with the valence of +4 expected for carbon. For hydrogen in palladium[2] $Q_e \cong 1$, as expected. Other systems have given Q_e values that are not so nearly consistent with the valence. The usual interpretation of such discrepancies is the existence of electron momentum transfer to the jumping ion. This extra force from the conduction electrons is exerted in addition to the motion of the charged ion in the electric field. Such a momentum transfer alters Q_e from the expected value of the solute valence.

Similar to Eq. 2.98, a steady state concentration gradient is expected for a temperature gradient. In that instance, the steady state gradient is given by

$$\frac{\partial \ln C}{\partial x} = -\frac{Q \nabla T}{K_B T^2}$$

or $\qquad\qquad\qquad\qquad\qquad\qquad\qquad\qquad$ 2.99

$$\frac{\partial \ln C}{\partial 1/T} = \frac{Q}{K_B T}$$

when ∇T is constant. Measurements of the steady state or Soret gradient, after its founder, yield values of Q. For example, for carbon in alpha iron, Q was measured[3] to be -23 kcal. For hydrogen in alpha zirconium, Q was measured[4] to be $+6$ kcal/mole.

Measurements showing clearly that the chemical potential gradient is the driving force for diffusion and not a concentration gradient were performed by Darken[5]. He welded two iron bars together, the first had a solute concentration of 0.4% C; the second had 0.4% C and 4% Si. Thus the carbon concentration gradient was initially zero and the concentration itself was initially continuous. After the diffusion anneal the carbon had assumed a discontinuity in concentration at the weld interface. The chemical potential of carbon at the weld interface was shown to be continuous. There was, therefore, a diffusive flux causing the chemical potential to approach a continuous distribution at the expense of continuity of carbon concentration. The silicon does not diffuse nearly as fast as the carbon because it is a substitutional solute, and it maintained a large difference in concentration between the two welded bars throughout the experiment. In fact the silicon was nearly immobile relative to the carbon. Eventually, one expects the silicon would homogenize in the bar and the carbon would follow.

REFERENCES

1. Dayal and L. Darken, *Trans. AIME* **188,** 1156 (1950).
2. Wagner and G. Heller, *Z. Physik Chem.* **46B,** 242 (1940).
3. P. Shewmon, *Acta Met.* **8,** 606 (1960).
4. A. Sawatsky, *J. Nuci. Mater.* **2,** 321 (1960).
5. L. Darken, *Trans. AIME* **180,** 430 (1949).

CHAPTER 3

SELF-DIFFUSION OF SUBSTITUTIONAL SOLUTES BY A VACANCY MECHANISM IN DILUTE ALLOYS

VACANCY EQUILIBRIUM IN PURE METALS

As the temperature of a pure metal crystal is raised from absolute zero, fluctuations in the kinetic energy of surface atoms become of sufficient magnitude that any particular surface atom can leave its atom site and move to another surface position. Periodically, the atoms on the second layer of atom sites will view an empty atom site on the first layer. Another fluctuation permits a second layer atom to jump to the first layer leaving a vacant atom site on the second layer. When dislocations or grain boundaries intersect the surface, they present energetically preferential locations for such a sequence of events. The defect so formed is then free to wander throughout the crystal, and the defect's neighbors can then jump into the empty site.

The number of defects that are formed in a crystal of N atoms is dependent on the temperature and the change in the Gibbs' free energy of the crystal when a single vacancy is formed. For example, if the sum in Eq. 2.9 is extended over the atoms whose frequencies and binding potentials are influenced by a single vacancy, then one expects the energy of formation of a single vacancy to follow (see Eq. 2.9).

$$\Delta G_v = \sum_i \Delta A_i + \Delta(P\underline{V}_i) \qquad 3.1$$

63

If the number of defects, n, is very small in comparison to the number of atoms N, then to a good approximation the formation from energy for n defects is $n \Delta G_v$. This implicitly assumes that no divacancies or vacancy pairs form even though there may be a binding energy to hold vacancies together. The assumption works in a great many cases since there are a large number of atom sites $(N+n)$ where the vacancies may wander without becoming neighbors. The mobility of a divacancy is so much greater than a single vacancy, however, that one must be cautious in neglecting divacancies as contributing to diffusion. This aspect of vacancies is quite complex to handle properly and is therefore omitted until Chapter 5.

The explicit free energy change associated with the formation of n vacancies is the sum of the formation free energy, $n \Delta G_v$, and the contribution from the configurational entropy, S_c. With the above assumption regarding the formation of vacancy clusters, the configurational entropy is found from the Boltzmann equation

$$S_c = K_B \ln \Omega_c \qquad\qquad 3.2$$

where Ω_c is the configurational degeneracy and equals $(N+n)!/(N!\,n!)$. Combining these, the crystal free energy, G, is given by

$$\exp -\frac{G}{K_B T} = \exp -\frac{G_0}{K_B T} \sum_n \Omega_c \exp -\frac{n \Delta G_v}{K_B T} \qquad\qquad 3.3$$

where G_0 contains the kinetic and potential energy contributions to the crystal at the temperature and pressure in question. With the frequency distribution problem in mind, G_0 would be calculated using appropriate pressure and volume modifications of Eq. 2.4. On introducing the function for Ω_c into Eq. 3.3., one may again use the maximum term method in this case to determine the most probable value of $n = n^*$. Thus with $\Omega_c = (N+n)!/(N!\,n!)$ as the usual approximation,

$$\exp\left(-\frac{G}{K_B T}\right) = \exp\left(-\frac{G_0}{K_B T}\right) \sum_n \frac{(N+n)!}{(N!\,n!)} \exp\left(-\frac{n \Delta G_v}{K_B T}\right)$$

$$= \exp\left(-\frac{G_0}{K_B T}\right) \sum_n t_n \qquad\qquad 3.4$$

The maximum term is found where

$$\frac{\partial [\{-G_0/K_B T\} + \ln t_n]}{\partial n}\Bigg|_{N,T} = 0 \qquad\qquad 3.5$$

at $n = n^*$. If one uses Sterling's approximation, $\ln X! = X \ln X - X$, then the solution to Eq. 3.5 is found when

$$\frac{n^*}{N+n^*} = \exp\left(-\frac{\Delta G_v}{K_B T}\right) \qquad 3.6$$

A more complete analysis is given in Appendix A, where divacancies are treated. Divacancy effects are discussed in a later chapter.

The summation in Eq. 3.4 can be reduced by the substitution of the maximum term for the series found in Eq. 3.6. When such a substitution is made, one may determine the Gibbs' free energy for the crystal including vacancies. Thus when one takes the maximum term as representative of the whole series in Eq. 3.4, one finds

$$G = G_0 + n^* \Delta G_v - K_B T \ln \frac{(N+n^*)!}{N!\, n^*!} \qquad 3.7$$

One may determine the chemical potential of the vacancies in the usual manner from Eq. 3.7. One finds the chemical potential of the vacancies $\mu_v = \partial G/\partial n^*|_{T,P}$ when equilibrium exists in the crystal. Thus at equilibrium

$$\mu_v = \Delta G_v + K_B T \frac{\partial \ln [n^*!/(N+n^*)!]}{\partial n^*}$$

$$= \Delta G_v + K_B T \ln \left(\frac{n^*}{N+n^*}\right) \qquad 3.8a$$

$$= 0 \qquad 3.8b$$

In Chapter 2, a discussion led to the diffusion flux being dependent on a chemical potential gradient. Equation 3.8 may cause one to assume that the chemical potential gradient of the vacancies does not lead to a contribution to the flux. This is not the case for the migration of substitutional solutes, and may be seen through the existence of a chemical potential gradient for the vacancies even though the vacancies are everywhere in equilibrium. In a temperature gradient, for example, if the vacancies are everywhere in equilibrium, Eq. 3.6 must be satisfied. The chemical potential gradient of the vacancies at constant temperature is found from Eq. 3.8 as

$$\nabla \mu_v|_T = \nabla(\Delta G_v) + K_B T \nabla \ln \left(\frac{n^*}{N+n^*}\right) \qquad 3.9a$$

$$= \nabla(\Delta G_v) - T \nabla \frac{\Delta G_v}{T}$$

$$= \frac{\Delta H_v}{T} \nabla T \qquad 3.9b$$

There is a chemical potential gradient for the vacancies because the temperature, as an implicit parameter in the vacancy concentration term, $n^*/(N+n^*)$, cause a vacancy concentration gradient in the crystal. Vacancy concentration gradients should be included in the phenomenological equations for the flux of substitutional solutes whenever they exist, even though the concentration of vacancies is everywhere in equilibrium. Equation 3.9 is derived assuming vacancies are everywhere in equilibrium. It has been recently shown that such an assumption is not valid for a temperature gradient in all systems. For a temperature gradient in a single crystal one must replace Eq. 3.9 with

$$\nabla \mu_v|_T = \frac{\gamma \, \Delta H_v}{T} \nabla T \qquad\qquad 3.10$$

where γ is a factor that lies between 0 and 1. Experimentally, $\gamma = 0.35$ for aluminum single crystals; the existence of this effect is discussed more thoroughly in the next chapter.

VACANCY-SUBSTITUTIONAL SOLUTE BINDING IN A DILUTE ALLOY

In a pure metal crystal, the introduction of a vacancy into the lattice is characterized by the removal of an ion to the surface. The free electrons, which may be viewed as a gas, will tend to fill the void so created. As a consequence there will be a negative charge associated with the vacant lattice site due to the electrons that replace the ion in the vicinity of the empty lattice site. When a substitutional solute has a valence different from the solvent, its ionic charge will interact with the electrons that fill the vacancy; this will give a coulombic attraction or repulsion of the solute ion with the vacancy. As an example, for monovalent solvent and divalent solute metal ions, the divalent solute ion will have a charge of $+2$, whereas the solvent has a charge of $+1$. The excess positive charge, $+1$, in the solute will attract a vacancy in the solvent. Thus, according to Lazarus, there will be an attractive energy holding the vacancy at a nearest-neighbor lattice site to the solute. Also, when the solute attempts to jump into the vacancy, the same coulombic interaction will decrease the amount of energy required for this action. Consequently, the amount of energy to form a vacancy at a site nearest neighbor to a solute will be different from the energy needed to form the vacancy in the absence of the solute. Also, the amount of energy for a solvent to jump into a vacancy will be different from that for a solute to jump into the vacancy.

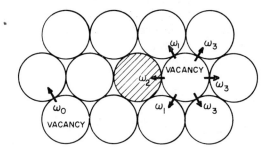

Figure 3.1 Illustration of vacancy-solute jump frequencies on (111) plane in an fcc lattice (solute is cross-hatched).

Such a physical idea may be described by letting the vacancy jump frequencies in the immediate vicinity of a solute ion in a very dilute binary alloy differ from the frequency in the absence of the solute. To this end, let w_0 be the jump frequency of a vacancy into a solvent atom far removed from any solute. In a very dilute solution, the solute atoms will be isolated from one another. Thus the above can be reflected by differing vacancy jump frequencies around a single solute in the lattice. Let w_1 be the exchange frequency of a vacancy with a solvent atom when both the vacancy and solvent are nearest neighbors to the solute, and w_2 be the frequency that a vacancy jumps into the solute position from a nearest neighbor site. Define w_3 to be the frequency with which a vacancy that is a nearest neighbor to a solute dissociates by jumping into a second neighbor solvent ion. Let w_4 be the frequency for the reverse jump from w_3; the vacancy becomes a nearest neighbor to the solute. (See Fig. 3.1.)

One may view the formation of a solute-vacancy pair in a dilute alloy as a kind of chemical reaction if one defines the neighboring vacancy-solute pair as an entity different from the isolated species. To this end let C be the fraction of lattice sites on which a solute atom resides under the circumstance that it has a neighboring vacancy. Let V be the fraction of lattice sites containing vacancies that are not nearest neighbors to solute ions. Also, I is defined as the fraction of lattice sites containing solute ions that have no neighboring vacancies, and let I_0 be the total fraction of lattice sites containing solute atoms. Assuming the solute concentration is small, the formation of solute clusters may be ignored. Then one has a conservation equation that

$$I_0 = I + C \qquad\qquad 3.11$$

Furthermore, the vacancy jumps permit the reaction

$$I + V \underset{R_2}{\overset{R_1}{\rightleftharpoons}} C \qquad\qquad 3.12$$

to occur, where R_1 and R_2 are the appropriate reaction rates. If two vacancy-solute clusters are also ignored, Eqs. 3.11 and 3.12 describe the only complexes that are formed. Equation 3.12 implies that the concentration of complexes is given by

$$\frac{dC}{dt} = R_1 IV - R_2 C \qquad\qquad 3.13$$

and at equilibrium, $dC/dt = 0$. Hence, at equilibrium

$$C = \frac{R_1 IV}{R_2} = \frac{VI_0(R_1/R_2)}{1 + V(R_1/R_2)} \qquad\qquad 3.14$$

The latter expression uses Eq. 3.11.

If Z is the number of nearest neighbors to the solute and M is the number of second neighbor sites that can be reached in one vacancy jump from a particular first neighbor site to the solute, then the reaction rates are

$$R_1 = ZMw_4$$

and

$$R_2 = Mw_3 \qquad\qquad 3.15$$

When $V \ll 1$ and w_4/w_3 is of the order of magnitude of unity (i.e., $w_4/w_3 \neq 10^{\pm 4}$ etc.), substituting Eq. 3.15 into Eq. 3.14 yields

$$C \cong \frac{VI_0 Zw_4}{w_3}$$

$$\cong ZI_0 \exp\left\{ -\frac{\Delta G_v - E_b}{K_B T} \right\} \qquad\qquad 3.16$$

where $w_4/w_3 = \exp[E_b/K_B T]$; E_b is the solute-vacancy binding energy.

Hence when the jump frequencies w_4 and w_3 differ, the concentration of vacancies at a solute departs from that expected for a particular solvent atom. This can easily be seen since the fraction of solute atoms that have associated vacancies in their first neighbor shell is C/I_0. For a given solvent atom site, the fraction of solvent atoms that have a neighboring vacancy is just ZV. Thus one may compare ZV to $C/I_0 = ZVw_4/w_3$. This serves to alter the total concentration of vacancies in the alloy. The fraction of lattice sites containing vacancies is $n^*/(N+n^*)$, where N is the number of atoms. Then with V as the vacancies on $(1 - ZI_0)$ sites, and C as the balance, the fraction of empty lattice sites becomes concentration dependent and follows:

$$\frac{n^*}{(N+n^*)} = V\left\{ 1 - ZI_0 + ZI_0 \frac{w_4}{w_3} \right\} \qquad\qquad 3.17$$

and $V = \exp\{-\Delta G_v/K_B T\}$ from Eq. 3.6.

SELF-DIFFUSION BY A VACANCY MECHANISM

Self-diffusion in metals may be defined as the diffusion of a tracer atom in a homogeneous alloy in the absence of an applied field. In such a situation the tracer flux is given by an equation similar to Eq. 1.6 with no velocity term due to an applied force. Thus the flux of component K will be given by

$$J_K = -D_K^* \frac{\partial C_K}{\partial x} \qquad 3.18$$

All of the information on the rate of the tracer motion is contained in the tracer self-diffusion coefficient D_K^*. Since the velocity term is zero, and since the tracer velocity is related to the average x displacement $\langle x_n(t_n) \rangle$ for n tracer jumps with nonzero x projection which occur in time t_n, $\langle x_n(t_n) \rangle = 0$ on the average. In what follows, it is shown that $\langle x_n^2 \rangle \neq 0$, and in fact the tracer self-diffusion coefficient is found from

$$D_K^* = \frac{\langle x_n^2 \rangle}{2 t_n} \qquad 3.19$$

One may easily justify Eq. 3.19 as follows: Equation 1.34 gives the solution to the motion of a tracer in an applied field, from a point source of strength Q_0, at the origin at time $t = 0$. When the field is zero (i.e., $\alpha = 0$), the tracer concentration given in Eq. 1.34 is

$$C_K(x, t) = \frac{Q_0}{\sqrt{\pi D_K^* t}} \exp \left\{ -\frac{x^2}{4 D_K^* t} \right\} \qquad 3.20$$

This means that the probability of finding a tracer atom at a position x at time t is $C_K(x, t)/2Q_0$. Letting $P_K(x, t)$ be this probability

$$P_K(x, t) = \frac{1}{2\sqrt{\pi D_K^* t}} \exp \left\{ -\frac{x^2}{4 D_K^* t} \right\} \qquad 3.21$$

One may calculate $\langle x^2 \rangle$ from this equation as follows:

$$\langle x^2 \rangle = \int_{-\infty}^{\infty} x^2 P_K(x, t) \, dx$$

$$= \int_{-\infty}^{\infty} \frac{x^2}{2\sqrt{\pi D_K^* t}} \exp \left\{ -\frac{x^2}{4 D_K^* t} \right\} dx \qquad 3.22$$

Now the integrals

$$\int_{-\infty}^{\infty} x^2 \exp(-ax^2)\, dx = \sqrt{\frac{\pi}{4a^3}} \tag{3.23a}$$

$$\int_{-\infty}^{\infty} \exp(-ax^2)\, dx = \sqrt{\frac{\pi}{a}} \tag{3.23b}$$

Using Eq. 3.23b, one may confirm that

$$\int_{-\infty}^{\infty} P(x, t)\, dx = 1$$

Furthermore, using Eq. 3.23a to evaluate Eq. 3.22, one finds that

$$\langle x^2 \rangle = 2D_K^* t$$

which is identical to Eq. 3.19.

For simplicity in what follows it is assumed that we are dealing with a cubic crystal oriented in such a manner that the X axis is perpendicular to (100) planes. This simplification introduces a single x jump distance. Thus the x distance of any tracer jump with nonzero x projection may be taken as b. Since the diffusion coefficient is isotropic, as shown in Chapter 1, this crystal orientation is as good as any other; it also simplifies the calculation.

To calculate $\langle x_n^2 \rangle$, one sums the x distance for n jumps. That is

$$x_n = \sum_{i=1}^{n} x_i$$

Then,

$$x_n^2 = \left(\sum_{i=1}^{n} x_i \right)\left(\sum_{i=1}^{n} x_i \right) \tag{3.24}$$

$$= x_1 \cdot x_1 + x_1 \cdot x_2 + x_1 \cdot x_3 + \cdots x_1 \cdot x_n$$
$$+ x_2 \cdot x_1 + x_2 \cdot x_2 + x_2 \cdot x_3 + \cdots x_2 \cdot x_n$$
$$+ \cdots \cdots \cdots \cdots \cdots \cdots \cdots \cdots \cdots$$
$$+ x_n \cdot x_1 + x_n \cdot x_2 + \cdots \cdots \cdots x_n \cdot x_n$$

The latter form for Eq. 3.24 suggests that one may group the terms along the diagonal and write the others as addition terms as follows:

$$x_n^2 = \sum_{i=1}^{n} x_i \cdot x_i + 2 \sum_{i=1}^{n-1} x_i x_{i+1} + 2 \sum_{i=1}^{n-2} x_i x_{i+2} + \cdots$$

$$= \sum_{i=1}^{n} x_i^2 + 2 \sum_{j=1}^{n-1} \sum_{i=1}^{n-j} x_i x_{i+j} \tag{3.25}$$

Since the X projection of any atomic jump is $\pm b$, and since only the nonzero terms in Eq. 3.25 are of interest, one knows that

$$\sum_{i=1}^{n} x_i^2 = nb^2$$

Substituting Eq. 3.26 into Eq. 3.25, one may write Eq. 3.25 in the form

$$x_n^2 = nb^2 \left\{ 1 + \frac{2}{n} \sum_{j=1}^{n-1} \sum_{i=1}^{n-j} \frac{x_i x_{i+j}}{b^2} \right\} \tag{3.26}$$

The diffusion coefficient is then related to Eq. 3.26 through Eq. 3.19 as follows:

$$D_K^* = \lim_{n \to \infty} \frac{\langle x_n^2 \rangle}{2 t_n}$$

$$= \lim_{n \to \infty} \frac{nb^2}{2 t_n} \left\{ 1 + \frac{2}{n} \sum_{j=1}^{n-1} \sum_{i=1}^{n-j} \frac{\langle x_i x_{i+j} \rangle}{b^2} \right\} \tag{3.27}$$

where n is the number of jumps with nonzero x projection.

If one defines the average tracer jump frequency Γ as

$$\Gamma = \lim_{n \to \infty} \frac{n}{t_n} \tag{3.28}$$

then it is apparent that the diffusion coefficient may be found from

$$D_K^* = \frac{b^2 \Gamma}{2} \lim_{n \to \infty} \left\{ 1 + \frac{2}{n} \sum_{j=1}^{n-1} \sum_{i=1}^{n-j} \frac{\langle x_i x_{i+j} \rangle}{b^2} \right\}$$

$$= \frac{b^2 \Gamma}{2} f \tag{3.29}$$

Equation 3.29 serves as a definition for the correlation factor f. Therein, one finds

$$f = \lim_{n \to \infty} \left\{ 1 + \frac{2}{n} \sum_{j=1}^{n-1} \sum_{i=1}^{n-j} \frac{\langle x_i x_{i+j} \rangle}{b^2} \right\} \tag{3.30}$$

One may compare the results given in Eqs. 3.29 and 3.30 to those given in Eq. 2.31 for the situation in which D_K^* equals a constant and v_K^1 equals zero. Such a comparison leads one to the conclusion that $f = 1$ for the diffusion of an interstitial solute. For diffusion of a substitutional solute by a vacancy mechanism, it will be found that the correlation factor is not generally unity. In fact, it will be seen that one generally has $0 \le f \le 1$.

The origin of the correlation factor is very straightforward. For the motion of an interstitial solute, each jump is independent of the previous jump; that is, all solute jumps are independent. For a dilute substitutional alloy, all vacancy jumps are also independent. However, the jumps of a solvent or solute tracer atom are not independent, they are correlated. Once a tracer has jumped into a vacancy yielding an x migration of $\pm b$, the next most probable tracer jump is for the tracer to jump back into the same vacancy. Thus the most probable net x displacement for the two jump sequence is zero. This is the reason that $f \leq 1$. The correlation factor corrects for the situation in which an n jump sequence does not lead to $x_n^2 = nb^2$; the displacement is nearly always less than this value for a vacancy mechanism.

The calculation of the diffusion coefficient is now reduced to the determination of the average tracer jump frequency, Γ, and the correlation factor, f.

STEADY-STATE CALCULATION OF THE SOLUTE JUMP FREQUENCY

The steady-state calculation of the tracer jump frequency is easily determined using the time average concentration of vacancies around a solute tracer atom. For a given tracer to jump, one requires a vacancy at a neighboring site into which the tracer can jump with a nonzero x projection. For either body-centered or face-centered cubic crystals oriented so that the (100) planes are perpendicular to the x axis, there are eight possible vacancy sites on planes adjacent to the one containing the tracer that would give an x displacement of $\pm b$ when the tracer and vacancy exchange positions. The probability that one of these sites contains a vacancy is $C/(ZI_0)$, where C is found from Eq. 3.16. The frequency with which the tracer and vacancy exchange positions was previously defined as w_2. Thus for the jumps that give x displacements in fcc or bcc, one has

$$\Gamma = \frac{8C}{ZI_0} w_2$$

$$= 8Vw_2 \frac{w_4}{w_3} \qquad\qquad 3.31$$

for the tracer jump frequency.

The factor of 8 originates from the 8 sites around the tracer that could lead to a vacancy-tracer exchange of $\pm b$ along the x axis; and C/ZI_0 is the probability that any one of these sites is vacant. Combining Eq. 3.31 with Eq. 3.29, one finds

$$D_K^* = 4b^2 V w_2 \frac{w_4}{w_3} f \qquad\qquad 3.32a$$

$$= 4b^2 \nu_K f \exp\left\{ -\frac{(\Delta G_m + \Delta G_v - E_b)}{K_B T} \right\} \qquad\qquad 3.32b$$

As before, ν_K is the vibrational frequency of the tracer, and ΔG_v is the formation free energy of the vacancy in the pure solvent. The term ΔG_m is the migration free energy of the solute into the vacancy, that is,

$$w_2 = \nu_K \exp\left(-\frac{\Delta G_m}{K_B T} \right)$$

and E_b is the binding energy of the vacancy to the tracer.

Generally for solute tracer diffusion in the alloy, f is expected to be somewhat temperature dependent. Experimentally, such a dependence is usually minor compared to the exponential dependences given in Eq. 3.32b. As a consequence, one can plot diffusion coefficients on a graph of $\ln D$ versus $1/T$. The slope of the line on such a graph is expected from Eq. 3.32b to be

$$\frac{\partial \ln D_K^*}{\partial 1/T} = -\frac{(\Delta H_v + \Delta H_m - E_b)}{K_B} \qquad\qquad 3.33$$

assuming f is constant. In Eq. 3.33 ΔH_v is the enthalpy of vacancy formation, and ΔH_m is the enthalpy of the vacancy solute tracer exchange.

For solvent tracer self-diffusion in the pure metal, E_b is zero since $w_4 = w_3 = w_2 = w_1 = w_0$ in this case. Furthermore, the correlation factor is a constant and hence the slope indicated by Eq. 3.33 gives the vacancy migration and formation energies. One can measure ΔH_v and ΔH_m by independent means and, within experimental error, there is excellent agreement between the measured ΔH_m and the measured ΔH_v, which sum to measured $(\partial \ln D_K^*)/(\partial 1/T)$ for the solvent tracer since in that instance

$$\frac{\partial \ln D_K^*}{\partial 1/T} = -\left(\frac{\Delta H_v + \Delta H_m}{K_B} \right)$$

DETERMINATION OF THE CORRELATION FACTOR

To determine the expression for the correlation factor, one proceeds to the limit described in Eq. 3.30 as follows: The major simplification of this equation is to note that $\langle x_i x_{i+j} \rangle = \langle x_\ell x_{\ell+j} \rangle$ for all ℓ. This comes about because the exchanges between a single vacancy and a single solute tracer are indistinguishable except for orientation. Since they are indistinguishable, the probability of two successive jumps in the same direction is independent of which jump is considered the first jump. Such a situation would also imply that the probability of two successive jumps in opposite directions is independent of which jump is considered to be the first jump. Thus the probability that the first jump is in the $+x$ direction and the second jump is in the $-x$ direction would be the same probability of the first jump being in the $-x$ direction and the second jump being in the $+x$ direction. Similarly, the probability of the first jump being a $+x$ jump followed by a second $-x$ jump is the same as the second jump being a $-x$ followed by a third being a $+x$ jump.

As a consequence of this simplification, $\langle x_i x_{i+j} \rangle = \langle x_1 x_{1+j} \rangle$ so that all the terms in the sum over i in Eq. 3.30 are identical, and the series contains $(n-j)$ identical terms. Thus one may write

$$f = 1 + 2 \lim_{n \to \infty} \sum_{j=1}^{n-1} \frac{(n-j)}{n} \frac{\langle x_1 x_{1+j} \rangle}{b^2} \qquad 3.34$$

When the limit is $n \to \infty$, and $|x_1 x_{1+j}| = b^2$, one concludes that

$$f = 1 + 2 \sum_{j=1}^{\infty} \frac{\langle x_1 x_{1+j} \rangle}{b^2} \qquad 3.35$$

The situation that $\langle x_i x_{i+j} \rangle = \langle x_\ell x_{\ell+j} \rangle$ is not always true. A specific example of such a situation is diffusion by vacancy pairs. Detailed discussion of the complications that arise, and examples of problems where this assumption breaks down, are discussed in Chapter 5.

Suppose that one defines $P_{11}^{(2)}$ as the probability that the second tracer jump is parallel, 11, in the x direction to the first tracer jump, and $P_a^{(2)}$ as the probability that the second jump is antiparallel, a, to the first tracer jump. For uncorrelated tracer motion, $P_{11}^{(2)} = P_a^{(2)}$ and $\langle x_1 x_2 \rangle = 0$. To calculate $\langle x_1 x_2 \rangle$, one sums the x distance for the two cases just mentioned times the appropriate probability. Thus

$$\langle x_1 x_2 \rangle = |x_1|\{(+|x_2|)P_{11}^{(2)} + (-|x_2|)P_a^{(2)}\}$$
$$= b^2\{P_{11}^{(2)} - P_a^{(2)}\} \qquad 3.36$$

This is the first term in the series within Eq. 3.35.

Similarly, one may define $P_{11}^{(\ell)}$ as the probability that the ℓth tracer jump is parallel to the first tracer jump, and the probability $P_a^{(\ell)}$ is the same for the antiparallel ℓ^{th} jump related to the first jump. One may therefore use these terms to find $\langle x_1 x_\ell \rangle$; $\ell > 1$ as indicated in Eq. 3.36

$$\langle x_1 x_\ell \rangle = |x_1| \{ (+|x_\ell|) P_{11}^{(\ell)} + (-|x_\ell| P_a^{(\ell)}) \}$$

$$= b^2 \{ P_{11}^{(\ell)} - P_a^{(\ell)} \} \qquad 3.37$$

Consider the case that $\ell = 3$; a sequence of three jumps has occurred such that the last two jumps are related to the first jump. There are four possible situations for the relation between the last two jumps: (a) two parallel jumps, (b) two antiparallel jumps, (c) a parallel-antiparallel pair, and (d) an antiparallel-parallel pair. This means that: (a) x_2 is parallel to x_1, and x_3 is parallel to x_2; (b) x_2 is antiparallel to x_1, and x_3 is antiparallel to x_2; (c) x_2 is parallel to x_1, and x_3 is antiparallel to x_2; and (d) x_2 is antiparallel to x_1, and x_3 is parallel to x_2. In cases (a) and (b) the x_3 jump is parallel to the x_1 jump, whereas in cases (c) and (d) they are antiparallel. Hence

$$P_{11}^{(3)} = [P_{11}^{(2)}]^2 + [P_a^{(2)}]^2$$

$$P_a^{(3)} = 2 P_{11}^{(2)} P_a^{(2)} \qquad 3.38$$

Finally, consider the situation in which: (a) the $(\ell-1)^{st}$ jump is parallel to the first jump, and the ℓth jump is parallel to the $(\ell-1)^{st}$ jump; (b) the $(\ell-1)^{st}$ jump is antiparallel to the first jump, and the ℓth jump is antiparallel to the $(\ell-1)^{st}$ jump; (c) the $(\ell-1)^{st}$ jump is parallel to the first jump, and the ℓth jump is antiparallel to the $(\ell-1)^{st}$ jump; and (d) the $(\ell-1)^{st}$ jump is antiparallel to the first jump, and the ℓth jump is parallel to the $(\ell-1)^{st}$ jump. From (a) and (b), the ℓth jump is parallel to the first jump. Hence

$$P_{11}^{(\ell)} = P_{11}^{(\ell-1)} P_{11}^{(2)} + P_a^{(\ell-1)} P_a^{(2)} \qquad 3.39$$

Similarly, for (c) and (d), one finds

$$P_a^{(\ell)} = P_{11}^{(\ell-1)} P_a^{(2)} + P_a^{(\ell-1)} P_{11}^{(2)} \qquad 3.40$$

Substituting Eqs. 3.39 and 3.40 into Eq. 3.37,

$$\langle x_1 x_\ell \rangle = b^2 \{ P_{11}^{(\ell-1)} P_{11}^{(2)} + P_a^{(\ell-1)} P_a^{(2)} - P_{11}^{(\ell-1)} P_a^{(2)} - P_a^{(\ell-1)} P_{11}^{(2)} \}$$

$$= b^2 \{ [P_{11}^{(\ell-1)} - P_a^{(\ell-1)}][P_{11}^{(2)} - P_a^{(2)}] \} \qquad 3.41$$

Thus one evidently has

$$\frac{\langle x_1 x_2 \rangle}{b^2} = \frac{\langle x_1 x_{\ell-1} \rangle}{b^2} \frac{\langle x_1 x_2 \rangle}{b^2}$$

$$= \frac{\langle x_1 x_{\ell-2} \rangle}{b^2} \frac{\langle x_1 x_2 \rangle}{b^2} \frac{\langle x_1 x_2 \rangle}{b^2}$$

$$= \left\{ \frac{\langle x_1 x_2 \rangle}{b^2} \right\}^{\ell-1} = \{ P_{11}^{(2)} - P_a^{(2)} \}^{\ell-1} \qquad 3.42a$$

$$= \frac{\langle x_1 x_{1+(\ell-1)} \rangle}{b^2} \qquad 3.42b$$

Hence from Eq. 3.42

$$\frac{\langle x_1 x_{1+j} \rangle}{b^2} = \left[\frac{\langle x_1 x_2 \rangle}{b^2} \right]^j = [P_{11}^{(2)} - P_a^{(2)}]^j \qquad 3.43$$

Substituting Eq. 3.43 into Eq. 3.35 yields

$$
\begin{aligned}
f &= 1 + 2 \sum_{j=1}^{\infty} \left[\frac{\langle x_1 x_2 \rangle}{b^2} \right]^j \\
&= \frac{1 + \langle x_1 x_2 \rangle / b^2}{1 - \langle x_1 x_2 \rangle / b^2} \\
&= \frac{1 + P_{11}^{(2)} - P_a^{(2)}}{1 + P_a^{(2)} - P_{11}^{(2)}}
\end{aligned}
\qquad 3.44
$$

Since the first result has only the second jump as a superscript, the superscript will henceforth be omitted. Thus one writes

$$f = \frac{1 + P_{11} - P_a}{1 + P_a - P_{11}} \qquad 3.45$$

EVALUATION OF P_{11} AND P_a FOR AN fcc CRYSTAL

When there is no applied field, it is reasonable to assume that the motion of a vacancy is random in a dilute alloy. In the vicinity of a solute atom, the vacancy jump frequency is changed to the values previously defined, w_i, $i = 1\text{–}4$, which differ from w_0. As a consequence the probability of the vacancy exchanging with a particular atom in an fcc lattice is no longer 1/12 for each of its nearest neighbors. As an example, if the vacancy is a nearest neighbor to the solute, it may dissociate at a rate w_3, and there are seven sites that can accomplish this. It may jump to another nearest neighbor site of the solute at a rate w_1, and there are four such sites. Also, it may jump into the solute site at a rate w_2. Thus the probability that the vacancy and solute exchange is

$$\frac{w_2}{w_2 + 4w_1 + 7w_3} \qquad 3.46$$

Such probabilities can be determined for any pair of sites in the crystal; if the sites are not nearest neighbors, the probability of an exchange in one vacancy jump would be zero.

Suppose one labels the sites in the crystal where a vacancy could be found relative to the position of the solute atom. One could consider the vacancy at site j prior to a jump and let A_{ij} be the probability that the vacancy jumps to site i in one vacancy jump. If by some means one identifies the probability that the vacancy is found on site j after $n-1$ jumps as $P_{n-1}^{(j)}$, then one may determine the probability that the vacancy is on site i after n jumps through

$$P_n^{(i)} = A_{ij}P_{n-1}^{(j)}$$

If one further considers all possible sites j, then the total probability of occupancy of site i after n jumps is

$$P_n^{(i)} = \sum_j A_{ij}P_{n-1}^{(j)} \qquad 3.47$$

To simplify Eq. 3.47, one writes $P_n^{(i)}$ as the elements of a column vector, P_n, and the elements A_{ij} as a matrix A. Then one has

$$P_n = AP_{n-1} \qquad 3.48$$

which is valid for all n.

Thus

$$P_1 = AP_0 \qquad 3.49$$

where P_0 is some particular initial vacancy configuration of interest. Furthermore,

$$P_2 = AP_1 = A^2P_0$$

and

$$P_n = A^nP_0 \qquad 3.50$$

by induction.

We wish to calculate the total probability that the vacancy is found in a particular configuration relative to the solute for any number of vacancy jumps from 0 to ∞ from an initial starting point described by P_0. This probability is an element of the column vector

$$P_T = \sum_{n=0}^{\infty} P_n \qquad 3.51$$

Hence if the vacancy position is initially at P_0, it is found at any other site in the set of configurations within A_{ij} to

$$P_T = \sum_{n=0}^{\infty} P_n = \sum_{n=0}^{\infty} A^nP_0 = (I-A)^{-1}P_0 \qquad 3.52$$

In Eq. 3.52, I is the N-dimensional unit matrix where N is the dimensionality of A, and P_T and P_0 are N-dimensional column vectors. Stated simply, the matrix A randomly walks the vacancy throughout the crystal; the matrix function $(I-A)^{-1}$ adds the probabilities of the vacancy jump for all possible jumps; and P_0 gives the initial vacancy position. Clearly, P_{11} and P_a can be calculated using simple variations on Eq. 3.52.

In practice, one wishes to use the smallest dimensions possible on the matrix A and the vectors P_n, P_T, and so forth. Crystalline symmetry provides the greatest possible reduction consistent with a given accuracy. That is, the accuracy of the calculation depends on the number of configurations one uses; infinite accuracy would require all possible vacancy-solute configurations in the crystal. Such a case is clearly impossible, so one uses a finite number of vacancy positions surrounding the solute.

For various crystalline orientations relative to the x axis, many vacancy-solute configurations are related through rotational symmetry about the x axis. As an example, consider the nearest neighbor sites surrounding a solute in an fcc crystal with the x axis parallel to the $\langle 100 \rangle$ direction. There are four sites to the left of the tracer on the adjacent (100) plane, four sites to the right, and four sites on the same plane as the tracer. One may group these sites into three classes; class 1 is to the left, class 2 is to the right, and class 3 is on the same plane. Now each site in a class can be rotated into any other site in the same class through a 90° rotative about an axis through the tracer and parallel to the $\langle 100 \rangle$ direction. (See Fig. 3.2.) Furthermore the specific probability that a vacancy on a site in class 1 may jump to any other site is the same for all sites in class 1.

Since the above jump probabilities are the same, one may reduce the dimensionality of A, P_T, P_0 by considering the jump probabilities from members of a class of sites to members of another class of sites. A class of sites is defined by the configurations that are equivalent through an n-fold rotation about the x axis.

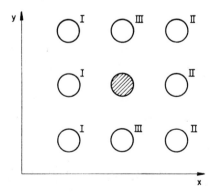

Figure 3.2 (100) Plane projection of vacancy-solute configurations in an fcc lattice (solute atom is cross-hatched, roman numerals refer to configuration classes).

As a consequence of using classes of sites, one must redefine the elements of the matrix A to be consistent with the such use. Thus in practice, A_{ij} is the probability that the defect will move in one jump from any site in class j to a particular site within class i. Also, P_0 is then the probability that the vacancy is initially in a particular site within a certain class. The elements of A_{ij} are determined by inspection.

For the three classes of sites surrounding the solute in an fcc crystal previously described, consider the following matrix elements for vacancy jumps. The jump probability from a particular site in class 1 to another neighboring particular site in class 1 is

$$\frac{w_1}{w_2 + 4w_1 + 7w_3}$$

Furthermore, there are two sites in class 1 that can reach a particular site in class 1 with one vacancy jump. Thus

$$A_{11} = \frac{2w_1}{w_2 + 4w_1 + 7w_3} = \frac{2w_1}{d} \qquad 3.53$$

where $d = w_2 + 4w_1 + 7w_3$. As a second example, consider A_{33}. The probability of going from a particular site in class 3 to another site in class 3 is w_2/d. There is only one site in class 3 that can jump to a particular site in class 3 with one vacancy jump. Hence

$$A_{33} = \frac{w_2}{d} \qquad 3.54$$

For reasons explained later, the solute-vacancy exchanges with nonzero x projection are excluded from A. Hence $A_{21} = A_{12} = 0$. Otherwise, the matrix A is the following

$$A = \begin{pmatrix} \dfrac{2w_1}{d} & 0 & \dfrac{2w_1}{d} \\ 0 & \dfrac{2w_1}{d} & \dfrac{2w_1}{d} \\ \dfrac{2w_1}{d} & \dfrac{2w_1}{d} & \dfrac{w_2}{d} \end{pmatrix} \qquad 3.55$$

To calculate P_{11} and P_a, consider the second tracer jump as related in orientation to the first tracer jump. For P_a assume that the first tracer jump was in the positive x direction. An antiparallel jump would, therefore, be in the negative x direction. A tracer jump in the negative x direction occurs from class 1 sites, as previously described, since class 1 sites have the vacancy to the left of the tracer. Therefore, following tracer jump 1, the vacancy is in class 1 sites. The probability that it exists in a particular class 1 site is 1/4 since there are 4 sites in class 1. Thus to calculate the probability P_a, the vacancy must begin its motion with an initial probability distribution as follows:

$$P_0 = \frac{1}{4}\begin{pmatrix} 1 \\ 0 \\ 0 \end{pmatrix} \qquad 3.56$$

An antiparallel jump can occur with a probability of

$$\frac{w_2}{w_2 + 4w_1 + 7w_3}$$

from each of the four sites in class 1. Hence one defines the tracer transition probability as a row vector. The magnitude of the vector components are the jump probabilities from various configuration classes, β. Thus $Q(\beta)$ is a row vector that permits the vacancy and tracer to exchange from configuration β. For an antiparallel jump, the vacancy is in class 1. The jump probability for the antiparallel jump is therefore

$$Q(\beta) = \frac{4w_2}{w_2 + 4w_1 + 7w_3} (1, 0, 0) \qquad 3.57$$

when P_0 is given by Eq. 3.56. Upon combining Eqs. 3.52, 3.55, and 3.56, one gets the total probability, P_T, of the vacancy starting in class 1 and going to any other class in an infinite number of jumps. The scalar product of P_T with $Q(\beta)$ from Eq. 3.57 gives P_a. Hence

$$P_a = Q(\beta)(I - A)^{-1} P_0 \qquad 3.58$$

where specifically Eqs. 3.55, 3.56, and 3.57 are used in Eq. 3.58. It was previously mentioned that the matrix A excludes tracer jumps with nonzero X projection. If this were not the case, multiple tracer jumps would be present in Eq. 3.58, and P_a requires only one tracer jump, which is provided by $Q(\beta)$.

To calculate P_{11} one need only change the jump configuration, since if the first jump were in the position X direction, then the second jump would be in the positive X direction for P_{11}. The vacancy must be in configuration class 2 to make the parallel jump if it were in class 1 after the first jump. Thus letting the class 2 sites give β^I jumps,

$$P_{11} = Q(\beta^I)(I - A)^{-1} P_0 \qquad 3.59$$

The superscript I denotes the mirror image sites. Specifically, A and P_0

are given by Eqs. 3.55 and 3.56 but

$$Q(\beta^I) = \frac{4\,w_2}{w_2 + 4\,w_1 + 7\,w_3}(0, 1, 0) \qquad 3.60$$

Note that in Eq. 3.59 the vacancy starts its motion in class 1 sites, moves around the tracer to class 2 sites without jumping into the tracer, and then exchanges with the tracer one time.

Specifically, the results of the Eqs. 3.58 and 3.59 are

$$P_a = \frac{w_2(WX - 4\,w_1^2)}{W(WX - 8\,w_1^2)} \qquad 3.61a$$

$$P_{11} = \frac{4\,w_1^2 w_2}{W(WX - 8\,w_1^2)} \qquad 3.61b$$

where

$$W = d - 2\,w_1 = w_2 + 2\,w_1 + 7\,w_3$$

and

$$X = d - w_2 = 4\,w_1 + 7\,w_3$$

The correlation factor may be calculated using Eqs. 3.45 and 3.61. Thus from Eq. 3.61

$$P_{11} - P_a = -\frac{w_2}{W}$$

and

$$f = \frac{1 + (P_{11} - P_a)}{1 - (P_{11} - P_a)}$$

$$= \frac{2\,w_1 + 7\,w_3}{2\,w_1 + 2\,w_2 + 7\,w_3} \qquad 3.62$$

The accuracy of Eq. 3.62 may be improved by taking a larger number of configurations into the matrix A. This has been done by Manning[1] who finds that

$$f = \frac{2\,w_1 + 7Fw_3}{2\,w_1 + 2\,w_2 + 7Fw_3} \qquad 3.63$$

where, if one lets $w_4/w_0 = \theta$

$$F = 1 - \frac{1}{7}\left[\frac{10\theta^4 + 180.5\theta^3 + 927\theta^2 + 1341\theta}{2\theta^4 + 40.2\theta^3 + 254\theta^2 + 597\theta + 436}\right] \qquad 3.64$$

Note that when $\theta \gg 1$, $F = \frac{2}{7}$; when $\theta = 1$, $F = 5.15/7$; and when $\theta \ll 1$, $F = 1$. This gives a value for f, for self-diffusion of a solvent tracer in a pure metal, as $f = 0.781$ since $w_0 = w_1 = w_2 = w_3 = w_4$. On the other hand, if one evaluates Eq. 3.62 for these conditions, $f = \frac{9}{11} = 0.82$. The difference is about 6%, and illustrates the accuracy of the calculation giving P_a and P_{11} in Eq. 3.61. It is notable that the correlation factor is a constant in these conditions. Otherwise, since the jump frequencies are proportional to Boltzmann factors, the correlation factor in Eqs. 3.62 or 3.63 is temperature-dependent, which should influence the temperature dependence of D_K^*.

The self-diffusion coefficient for a dilute solute in an fcc crystal is found by combining Eq. 3.32 with Eq. 3.63. Thus

$$D_K^* = 4b^2 \frac{Vw_2w_4(2w_1 + 7Fw_3)}{w_3(2w_1 + 2w_2 + 7Fw_3)} \qquad 3.65$$

Perhaps the most interesting limiting case for the diffusion coefficient, in light of correlation effects, is the situation where $w_2 \gg w_i$, $i \neq 2$. Without correlations, one would intuitively expect to have a very large diffusion coefficient. This is not the case, as can be seen in Eq. 3.65. When W_2 becomes large, one finds

$$D_K^* = 4b^2 \frac{Vw_4}{w_3} (w_1 + \tfrac{7}{2}Fw_3) \qquad 3.66$$

Herein, the rate of solute diffusion is independent of the solute jump frequency w_2. The diffusivity becomes related to the arrival rate of the vacancy, w_4, the rate of vacancy dissociation from the solute, w_3, and the rate that the vacancy can encircle the solute, w_1, without making solute jumps. Interestingly, a plot of $\ln D_K^*$ versus $1/T$ for Eq. 3.66 might be interpreted as nearly self-diffusion of the solvent if $w_4 \cong w_3 \cong w_1 \cong w_0$. Such a plot is the usual method to interpret experimental data, and this example illustrates the caution one should use. In this particular instance, other experimental information would be needed to show that $w_2 \gg w_i$ for $i \neq 2$. One possibility is the measurement of the correlation factor which, for large w_2, $f \to 0$; the isotope effect is a convenient measurement to determine f.

RANDOM WALK CALCULATION OF THE JUMP FREQUENCY AND DIFFUSIVITY[2]

Vacancy equilibrium at a particular site in a crystal is an average value that reflects the random motion of the point defects. The diffusion coefficient for a vacancy mechanism is sensitive to the correlation factor, and the correlation factor reflects the net square distance a tracer moves by a single vacancy. Vacancy equilibrium at the tracer occurs as an average for the entire sequence of tracer-vacancy jumps until a single vacancy is randomized or finds a sink; then another vacancy comes along. Since the correlation factor averages the tracer motion over the sequence of vacancy exchanges, the calculation of the correlation factor must commence with the first vacancy-tracer exchange. Furthermore, the concentration of vacancies used to calculate the diffusion coefficient must reflect the concentration at the tracer for this first tracer jump, since the correlation factor and average number of jumps describe all subsequent jumps. This concentration of vacancies at the tracer, which permits the first jump, is not the equilibrium concentration unless the tracer and vacancy interchange positions but one time; the correlation factor is unity under such a circumstance. When the correlation factor is not unity, the concentration of vacancies at the tracer that permit the first jump is reduced from the equilibrium value; however, as described below, the average number of jumps becomes large and results in vacancy equilibrium. An explicit expression for the diffusion coefficient includes all of these events.

To follow the entire fluctuation leading to tracer diffusion and subsequent vacancy equilibrium at the tracer, one views the tracer at an arbitrary site in a cubic crystal with the X direction oriented perpendicular to (100) crystal planes. The solute concentration is small, and the tracer may be considered to be at some distance from defect sources and sinks. The simplest way to view the situation is for the tracer to be the center of a sphere whose surface contains a uniform distribution of sources and sinks. With a tracer jump distance of λ, the x projection of the tracer jump is b. The defect starts its motion at the source and migrates up to the tracer. Since one must average all subsequent returns of this particular defect to the tracer, it is important to specify that the defect has not previously arrived at a site where it can exchange with the tracer. Defect equilibrium will be assumed to exist on the surface of the sphere of sources. Defect equilibrium will be shown to be the average for sites inside the sphere for the entire path of the vacancy from its source to its sink.

In general, there will be a number of tracer-defect configurations that may lead to a single tracer jump with nonzero x projection. However, for the single vacancy-tracer mechanism there is only one, and its mirror image. When the defect is in a jump configuration, it is also in a configuration that would immediately follow a tracer jump in the negative x direction from a configuration α. From this position, the defect may exchange with the tracer in the $+x$ direction making a β jump. Consequently, one defines the rate of formation, $\nu_{\pi-\alpha}$, of the jump complex that immediately follows a jump of type α in the negative x direction, by a defect that has not previously been in any jump configuration. Such a defect must have originated its motion at a source and has not previously neighbored the tracer. Once the $-\bar{a}$ configuration is formed, several things could happen. Consequently, one must consider additional definitions. To this end, P_a is defined as the total effective probability of a jump in the x direction that is antiparallel, "a," to a previous jump $-\alpha$. The defect could also wander around the tracer to its other side without causing a tracer displacement in the $+x$ direction, and ultimately cause a tracer jump in the $-x$ direction. One, therefore, defines P_{11} as the effective probability that following an α tracer jump in the $-x$ direction, the next tracer defect exchange is of type β in the $-x$ direction, and would be parallel to the previous jump.

If the defect first arrives at the tracer by jumping into a configuration that follows a tracer jump of $-\alpha$, without previously exchanging with the tracer, and without arriving at any other configuration that could cause a tracer jump in the x direction, then only two possible tracer jumps could occur. The defect could cause a tracer displacement of: (1) $+b$ with an effective probability P_a, or the tracer could move a distance (2) $-b$ with a probability of P_{11}.

If one assumes that case (1) occurs, then the next tracer displacement could give a net displacement of: (3) $+2b$ with a probability P_{11}, or a net displacement of (4) $0b$ with a probability of P_a. On the other hand, if case (2) occurs, then the tracer could make a net displacement $-2b$ with a probability P_{11} or (6) $0b$ with a probability P_a.

The probability that (1) occurs and is followed by the defect migrating to a sink is evidently

$$P_a(1 - P_{11} - P_a) \qquad\qquad 3.67$$

The term in parenthesis is the probability that no further jumps occur, which implies that the β jump was the first and last tracer jump. On the other hand, process (2) could occur and be followed by the defect going to a sink. The effective probability for this is as follows

$$P_{11}(1 - P_{11} - P_a) \qquad\qquad 3.68$$

Consequently,

$$\nu_{\pi-\alpha}(P_{11} + P_a)(1 - P_{11} - P_a)$$

are the total number of jumps per unit time for a one jump sequence. Furthermore, the frequency of occurrence of tracer jump sequences that start as described above, and have exactly n jumps, is

$$\Gamma_n = \nu_{\pi-\alpha}(P_{11} + P_a)^n(1 - P_{11} - P_a) \qquad\qquad 3.69$$

Therefore, the total frequency Γ of jumps in these sequences is given by

$$\Gamma = \sum_{n=1}^{\infty} n\Gamma_n = \nu_{\pi-\alpha}(P_{11}+P_a)(1-P_{11}-P_a)^{-1} \qquad 3.70$$

With each jump providing an x-displacement of either $+b$ or $-b$, the diffusion coefficient D can be written as

$$D_K^* = \tfrac{1}{2}b^2 \nu f \qquad 3.71$$

where ν is the jump frequency for $+b$ jumps, and f is the correlation factor for the jumps. Since only half of the vacancies approach the tracer from the negative side, only half of the total jumps are included in the sequence appearing in Eq. 3.70. Thus since $+\alpha$ and $-\alpha$ jumps are equally likely to occur,

$$\nu = 2\Gamma \qquad 3.72$$

The correlation factor is

$$f = 1 + 2(P_{11}-P_a)(1-P_{11}+P_a)^{-1} \qquad 3.73$$

If one substitutes $P_{11}+P_a = P_1$ and $P_{11}-P_a = P_2$, then Eqs. 3.70–3.73 yield $f = (1+P_2)(1-P_2)^{-1}$ and

$$D_K^* = b^2 \nu_{\pi-\alpha} P_1(1-P_1)^{-1}(1+P_2)(1-P_2)^{-1} \qquad 3.74$$

To evaluate Eq. 3.74 and thereby the diffusivity, one defines a set of defect walk matrices such as A.

The first transition probability matrix of interest herein is A, representing M defect configurations; A is the same matrix used earlier. In addition, $P_0(\alpha)$ is an M-dimensional column vector of occupation probabilities following a tracer jump of type α. The term $P_0(\alpha^I)$ is the tracer jump from the image configuration. Furthermore, $Q(\beta)$ is the tracer transition probability for a β type jump written as a row vector, and $Q(\beta^I)$ is the reverse jump. With these definitions, one can follow a set of tracer-defect exchanges to determine the terms in Eq. 3.74. The terms in D are found in the following:

$$P_{\alpha\beta}^a = Q(\beta)(I-A)^{-1}P_0(\alpha^I)$$
$$P_{\alpha\beta}^{11} = Q(\beta)(I-A)^{-1}P_0(\alpha) \qquad 3.75$$

To evaluate $\nu_{\pi-\alpha}$ go to the set of defect sites that are one defect jump beyond the limits of the class of sites incorporated in A. These outer defect sites are assumed to be the source sites for the defect. Define $T(\gamma)$ as the number of defects per unit time jumping from a set of outer defect sites into a particular defect site γ, contained in the set of sites included in A. The term $T(\gamma)$ can be written as a column vector. Now $\nu_{\pi-\alpha}$ is the rate at which defects leave the outer defects sites and arrive at $-\alpha$, without exchanging with the tracer, and without arriving at any other site where a tracer exchange can occur. Hence from $T(\gamma)$, one finds the probability that the defect path will reach $-\alpha$; this probability is $H_{-\alpha\gamma}$:

$$\nu_{\pi-\alpha} = \sum_{\gamma} H_{-\alpha\gamma} T(\gamma) \qquad\qquad 3.76$$

Consider the class of defect sites, σ, one jump removed from a $-\alpha$ site, in a direction away from the tracer. One can write

$$H_{-\alpha\gamma} = \sum_{\sigma} H_{-\alpha\sigma} H_{\sigma\gamma} \qquad\qquad 3.77$$

The matrix with elements $H_{-\alpha\sigma}$ is defined as the occupation probability of a particular site, in $-\alpha$ times the number of sites in class $-\alpha$, times the jump probability from class σ. The matrix with elements $H_{\sigma\gamma}$ is found from the sum of the probabilities for all defect paths from γ to σ. For this purpose define a second defect walk matrix A_3. The elements A_{3ij} are the transition probabilities from the set of sites j into a particular site i. Here A_3 excludes all tracer jumps; and it excludes all defect jumps out of sites α and α^I. With this definition, the matrix H with elements $H_{-\alpha\gamma}$ is

$$H = (I - A_3)^{-1} \qquad\qquad 3.78$$

Since A_3 excludes defect jumps leaving sites α and α^I, the function $(I - A_3)^{-1}$ only permits the defect to arrive at α or α^I one time. This accomplishes the transition described in Eq. 3.77, incorporating all defect paths to the tracer in one closed form expression.   From Eqs. 3.76–3.78 the element $\nu_{\pi-\alpha}$ is found from

$$\nu_{\pi-\alpha} = H_{-\alpha}(I - A_3)^{-1} T \qquad\qquad 3.79$$

where $H_{-\alpha}$ is a row vector of occupation probabilities for sites in class $-\alpha$. The diffusion coefficient is found by combining Eqs. 3.74, 3.75, and 3.79.

A simple example of the above is the case of a dilute solute tracer in an fcc lattice which diffuses by a vacancy mechanism. In what follows one must invert matrices associated with defect transitions, and hand calculations become cumbersome for any matrix inversion beyond 3×3. A matrix of this size is sufficient to incorporate the three classes of nearest neighbor sites to a solute tracer diffusing perpendicular to (100) planes in a face-centered cubic alloy. The next approximation would include those sites that are second nearest neighbor to the tracer, and this would require inversion of a larger matrix.

For the purpose stated above one defines the previous three classes of vacancy sites. Class 1 is the four sites to the left of the tracer that could lead to a vacancy jump in the $+x$ direction. Class 2 is the four mirror image sites, and class 3 is the four sites on the same plane as the tracer.

Class 3 sites have the equilibrium concentration of vacancies when one permits the total evolution of the fluctuation to occur; that is, one lets the correlated sequence of tracer vacancy jumps to terminate. Thus we consider two aspects of the fluctuation on class 3 sites as follows: (a) the concentration of vacancies prior to the formation of an α complex, and (b) the concentration of vacancies including the sequence of correlated tracer jumps. The calculation of $v_{\pi-\alpha}$ reflects condition (a). By definition it deletes those vacancy paths into class 3 sites from class 1 and 2 sites and, as a consequence, the steady-state concentration of vacancies on class 3 sites for condition (a) is less than the equilibrium concentration.

To show the validity of vacancy equilibrium on class 3 sites for the entire fluctuation, case (b), one can calculate the vacancy concentration on class 3 sites using the matrix A_2, which includes tracer exchanges, as follows:

$$A_2 = \begin{pmatrix} \dfrac{2w_1}{d} & \dfrac{w_2}{d} & \dfrac{2w_1}{d} \\[2mm] \dfrac{w_2}{d} & \dfrac{2w_1}{d} & \dfrac{2w_1}{d} \\[2mm] \dfrac{2w_1}{d} & \dfrac{2w_1}{d} & \dfrac{w_2}{d} \end{pmatrix} \qquad 3.80$$

where $d = w_2 + 4w_1 + 7w_3$.

With Eq. 3.80 one can calculate the rate of arrival of vacancies at a particular class 3 site from the source, which are the second neighbor sites to the tracer for this approximation, as follows;

$$7w_4 V\{[A_2(I-A_2)^{-1}]_{31} + [A_2(I-A_2)^{-1}]_{32} + [A_2(I-A_2)^{-1}]_{33} + I_{33}\} \quad 3.81$$

where V is the vacancy concentration at equilibrium at the source. The matrix subscripts denote the elements; these are the net transition probabilities to class 3 sites from class 1, class 2, class 3, and directly from the source, respectively. The rate at which vacancies leave class 3 sites is the steady-state concentration, N_V^3, times the total jump rate of $w_2 + 4w_1 + 7w_3 = d$. Equating these two rates, one has

$$N_V^3 d = 7w_4 V\{[(I-A_2)^{-1} - I]_{31} + (I-A_2)^{-1} - I]_{32} + (I-A_2)^{-1}_{33} \quad 3.82$$

Solving Eq. 3.82 with A_3 given in Eq. 3.80, one finds the expected result for case b.

$$N_V{}^3 = \frac{w_4 V}{w_3} \qquad\qquad 3.83$$

Equation 3.83 results from an explicit demonstration that vacancy equilibrium at a particular atom is a result of random vacancy motion. The concentration at the solute tracer is zero over a large percentage of time, and then the random motion of a vacancy starting from its source moves it into the vicinity of the tracer. While the vacancy is in the neighborhood of the tracer it can cause a sequence of tracer jumps. When the vacancy finally leaves the tracer, the average of the time the tracer had a neighboring vacancy is recognized as the equilibrium value. Such a set of events is expressed in Eqs. 3.82 and 3.83.

When vacancy paths from the source to an atom exclude vacancy sites, the concentration of vacancies will become a nonequilibrium value.

For the situation in which class 3 sites are at vacancy equilibrium, one can easily calculate $v_{\pi-\alpha}$. For each site in $-\bar{\alpha}$, there are seven sites where a vacancy can jump into the α site at a rate w_4. The vacancy concentration on each of these seven sites is the equilibrium concentration for the source, V. There are also two sites that can act as sources on the plane containing the tracer, class 3 sites, and are neighbors to the tracer. These sites are assumed to have the equilibrium vacancy concentration of Vw_4/w_3 and a jump to the $-\bar{\alpha}$ site at a rate of w_1. Thus one finds

$$v_{\pi-\alpha} = 4[7w_4 V + 2w_1 N_V{}^3] \qquad\qquad 3.84a$$

$$= 4\left[7w_4 V + 2w_1 \frac{Vw_4}{w_3}\right]$$

$$= 4\frac{w_4}{w_3}(7w_3 + 2w_1)V \qquad\qquad 3.84b$$

when $N_V{}^3$ is the general concentration of vacancies on class 3 sites, and is found from $N_V{}^3 = (w_4/w_3)V$ for equilibrium only, Eq. 3.84b. Otherwise, Eq. 3.84a is applicable in general even when $N_V{}^3$ is a nonequilibrium value as is needed in the real calculation of $v_{\pi-\alpha}$.

On the other hand, if one is uncertain whether class 3 sites maintain vacancy equilibrium, one may use Eq. 3.79 to calculate $v_{\pi-\alpha}$. Alternatively, one may calculate the average concentration of vacancies on class 3 sites and follow the procedure given in Eq. 3.82. The latter method is shown first; both methods give the same results.

For case (a) and, for the correct calculation of $v_{\pi-\alpha}$ for this procedure, a different A_3 matrix must be used in Eq. 3.82. This new A_3 matrix omits all transitions from class 1 and 2 sites because the vacancy has yet to arrive in class 1 or 2 sites. Hence

$$A_3 = \begin{pmatrix} 0 & 0 & \dfrac{2w}{d} \\ 0 & 0 & \dfrac{2w_1}{d} \\ 0 & 0 & \dfrac{w_2}{d} \end{pmatrix} \qquad 3.85$$

When one substitutes Eq. 3.85 into Eq. 3.82, one finds

$$N_V^3 = \frac{7w_4 V}{4w_1 + 7w_3} \qquad 3.86$$

and this result may be compared with Eq. 3.83. Substituting Eq. 3.86 into Eq. 3.84a, the correct rate of arrival at diffusion sites, $-\bar{\alpha}$, is found to be

$$v_{\pi-\alpha} = 4(7w_4)\left(\frac{6w_1 + 7w_3}{4w_1 + 7w_3}\right)V \qquad 3.87$$

The final portion of the calculation of the diffusion coefficient requires the determination of P_{11} and P_a, Eq. 3.75. For this calculation one needs the matrix A given by Eq. 3.55.

Also, the matrices Q, $P_0(\alpha^I)$ and $P_0(\alpha)$ are needed. By definition, one has

$$Q(\beta) = \frac{4w_2}{d}(1 \quad 0 \quad 0)$$

$$P_0(\alpha^I) = \frac{1}{4}\begin{pmatrix} 1 \\ 0 \\ 0 \end{pmatrix}$$

and

$$P_0(\alpha) = \frac{1}{4}\begin{pmatrix} 0 \\ 1 \\ 0 \end{pmatrix} \qquad 3.88$$

as given earlier.

Substituting Eqs. 3.55 and 3.88 into Eq. 3.75 yields

$$P_a = \frac{(WX - 4w_1^2)w_2}{W(WX - 8w_1^2)} \qquad 3.61a$$

$$P_{11} = \frac{4w_1^2 w_2}{W(WX - 8w_1^2)} \qquad 3.61b$$

where $W = d - 2w_1$ and $X = d - w_2$.

The diffusion coefficient is then found by substituting Eqs. 3.87 and 3.61 into Eq. 3.74. With these substitutions, one finds the diffusion coefficient is given by the usual expression

$$D_K^* = 4b^2 \frac{w_2 w_4}{w_3} Vf$$

where f in this approximation is $(2w_1 + 7w_3)/(2w_1 + 2w_2 + 7w_3)$.

It is also important to calculate the rate of arrival of vacancies at the tracer $\nu_{\pi-\alpha}$, using Eq. 3.79. To do this, we need only the vectors $H_{-\alpha}$ and T, as well as the matrix A_3 previously defined in Eq. 3.85. The row vector $H_{-\alpha}$ is the occupation probability for the class of defect sites following a tracer jump of $-\alpha$. In that instance, the vacancy is in class 1 sites. The probability of being in a particular class 1 site is 1. There are four sites in class 1; hence

$$H_{-\alpha} = 4(1, 0, 0) \qquad\qquad 3.89$$

The column vector T is the jump rate of vacancies into particular sites in all classes that can be reached in one vacancy jump from the source. Since in this approximation the source is the second neighbor sites to the tracer, all three tracer jump classes can be reached in one vacancy jump from the source. Any particular site in either class 1, 2, or 3 can be reached from seven source sites at a jump rate of w_4. Thus the rate of arrival at any site in the three classes is $7w_4$ times the concentration of vacancies at the source, which is V. Thus T is the following

$$T = 7w_4 V \begin{pmatrix} 1 \\ 1 \\ 1 \end{pmatrix} \qquad\qquad 3.90$$

Substituting Eqs. 3.85, 3.89, and 3.90 into Eq. 3.79, yields the same results as previously described in Eq. 3.87.

Equation 3.34 evaluates the correlation factor for the limit as the number of jumps becomes infinite. This seems appropriate for the average jump frequency evaluated using the steady state concentration of vacancies at the solute, Vw_4/w_3. The average number of jumps per vacancy, as calculated by Eq. 3.70, is obviously finite. This brings up the question of the correctness of the formulation of the correlation factor. Such a question can be answered by comparing a series expansion for $\langle x^2 \rangle$ for an arbitrary number of jumps with a similar series expansion evaluated by $D_K^*/\nu_{\pi-\alpha} = \langle x^2 \rangle$.

One assumes that a single vacancy has arrived at the solute tracer. The tracer can make a jump of $+b$ with a probability of P_a or a jump of $-b$

with a probability P_{11}. Thus for one jump the $\langle x^2 \rangle_1$ is

$$\langle x^2 \rangle_1 = b^2[(+1)^2(P_a) + (-1)^2(P_{11})](1 - P_{11} - P_a)$$
$$= b^2(P_a + P_{11})(1 - P_{11} - P_a) \qquad 3.91$$

The last term is the probability that only one jump occurred. The second jump can be a parallel jump so that the distance from the origin is $\pm 2b$ or an antiparallel jump with distance 0 since the tracer returns to the origin. Hence

$$\langle x^2 \rangle_2 = b^2(P_{11} + P_a)[4(P_{11})](1 - P_{11} - P_a) \qquad 3.92$$

For an extended sequence of jumps:

$$\frac{\langle x^2 \rangle}{b^2} = (P_{11} + P_a)[1 + 4P_{11} + 9(P_{11})^2$$

$$+ P_{11}P_a + P_aP_{11} + 16(P_{11})3$$
$$+ 4(P_{1i})^2P_a + 4P_{11}(P_a)^2 + 4P_a(P_{11})^2$$
$$+ 4(P_a)^2P_{11} + \cdots](1 - P_{11} - P_a) \qquad 3.93a$$
$$= (P_{11} + P_a)[1 + 3P_{11} + 5(P_{11})^2$$
$$- P_a - 3P_{11}P_a + P_aP_{11} + \cdots] \qquad 3.93b$$

For comparison, $D_K^*/(\nu_{\pi-\alpha}b^2)$ is the following

$$\frac{D_K^*}{b^2\nu_{\pi-\alpha}} = P_1(1 - P_1)^{-1}(1 + P_2)(1 - P_2)^{-1} \qquad 3.94$$

Inserting the value of $P_1 = P_{11} + P_a$ and $P_2 = P_{11} - P_a$, Eq. 3.94 becomes

$$\frac{D_K^*}{\nu_{\pi-\alpha}b^2} = (P_{11} + P_a)\left[\sum_{n=0}^{\infty}(P_{11} + P_a)^n\right]$$

$$\times (1 + P_{11} - P_a)\left[\sum_{n=0}^{\infty}(P_{11} - P_a)^n\right] \qquad 3.95a$$
$$= (P_{11} + P_a)[1 + P_{11} + P_a + (P_{11})^2$$
$$+ P_{11}P_a + P_aP_{11} + (P_a)^2 + \cdots]$$
$$\times (1 + P_{11} - P_a)[1 + P_{11} - P_a + P_{11}^2$$
$$- P_{11}P_a - P_aP_{11} + (P_a)^2 + \cdots] \qquad 3.95b$$
$$= (P_{11} + P_a)[1 + 3P_{11} - P_a + 5(P_{11})^2$$
$$- 3P_{11}P_a + P_aP_{11} + \cdots] \qquad 3.95c$$

It is notable, therefore, that the series Eqs. 3.93*b* and 3.95*c* are identical. This confirmation shows the validity of the results for the correlation factor as given by Eq. 3.34.

EXPERIMENTAL RESULTS FOR fcc CRYSTALS

The conclusive experimental evidence for the vacancy mechanism of diffusion is provided by the agreement between the activation energy for self-diffusion for a pure metal tracer and the sum of the vacancy formation and migration energy for the same system. In theory, the activation energy should be the sum of the vacancy formation and migration energies, since the correlation factor is a constant for tracer diffusion of the pure metal. One obtains this energy through the slope of a plot of ln D versus $1/T$ for experimental data. The most reliable value for the vacancy formation energy is provided by the dilation experiments of Simmons and Balluffi (see Table 3.1). The same information can be obtained by electrical resistivity measurements to find quenched in vacancies. In addition, the quenching studies provide data on the vacancy migration energy. Thus Table 3.1 presents some of the available data that compare these experiments. It is notable that the best sum of $\Delta H_v + \Delta H_m$ is in agreement with Q to within 0.02 e.v.

Table 3.1   Activation energy for self-diffusion
$$D = D_0 \exp(-Q/K_B T)$$

Element	ΔH_v (e.v.)	ΔH_m (e.v.)	$\Delta H_v + \Delta H_m$ (e.v.)	Q (e.v.)
Au	0.98[a]	0.82[a]	1.80	1.80[b]
	0.94[c]		1.76	
Ag	1.01[d]	0.83[d]	1.84	1.91[a]
	1.09[e]		1.92	
Al	0.79[f]	0.52[f]	1.31	1.25[g]
	0.75[c]		1.27	

[a] J. Bauerle and J. Kochler, *Phys. Rev.* **107**, 1493 (1957).
[b] S. Makin, A. Rowe, and A. LeClaire, *Proc. Phys. Soc.* **70B**, 545 (1957).
[c] R. Simmons and R. Balluffi, *Phys. Rev.* **125** (1962).
[d] S. Gertsriken and N. Novikov, *Phys. Met. Metallog.* **9**, 54 (1960).
[e] C. Tomizuka and E. Sonder, *Phys. Rev.* **103**, 1182 (1956).
[f] W. DeSorbo and D. Turnbull, *Acta Met.* **7**, 83 (1957).
[g] S. Fradin and T. Rowland, *Appl. Phys. Lett.* **11**, 207 (1967).

Table 3.2 Impurity diffusion in dilute alloys

$$D = D_0 \exp(-Q/K_B T)$$

Impurity	Ionic Charge ΔZ	Q (e.v.)	Reference
Impurities in Cu			
Zn	+1	1.99	a
Hg	+1	1.91	b
Ga	+2	2.00	b
As	+4	1.83	b
Sb	+4	1.83	c
Impurities in Ag			
Cd	+1	1.81	d
Hg	+1	1.66	e
Zn	+1	1.81	d
In	+2	1.76	e
Tl	+2	1.74	f
Sn	+3	1.71	e
Pb	+3	1.66	g
Sb	+4	1.66	h

[a] I. Hino, C. Tomizuka, and C. Wert, *Acta Met.* **5,** 41 (1957).

[b] C. Tomizuka, footnoted by D. Lazarus, *Solid State Physics*, Vol. 10, Academic Press, New York, 1960, p. 117.

[c] M. C. Inman and L. W. Barr, *Acta Met.* **8,** 112 (1960).

[d] A. Sawatsky and F. Jaumot, *Trans. AIME* **209,** 1207 (1956).

[e] C. Tomizuka and L. Slifkin, *Phys. Rev.* **96,** 610 (1954).

[f] R. Hoffman, *Acta Met.* **6,** 95 (1958).

[g] R. Hoffman, D. Turnbull, and E. Hart. *Acta Met.* **3,** 417 (1955).

[h] E. Sonder, L. Slifkin, and C. Tomizuka, *Phys. Rev.* **93,** 970 (1954).

For comparison, Table 3.2 presents the activation energy for self-diffusion of various impurities in copper and silver. Such an activation energy also contains the temperature dependence of the correlation factor. These results are presented to emphasize the dependence on the difference in the ionic charge of the impurity and the solvent; such a charge difference was mentioned earlier as providing a physical basis for the jump frequencies W_i, $i = 1$–4, to differ from W_0. From this model by Lazarus, Q should be a decreasing function of ΔZ, as presented in the table. With few exceptions, the data reflect such a dependence.

These examples totally exclude the possible contribution of divacancies to the self-diffusion in fcc metals. Divacancies are now thought by many researchers to be an important contribution to the diffusion process, particularly as one approaches the melting temperature. Appendix A gives the theoretical background for these researchers views. A kinetic analysis is also presented in Chapter 5, and a computer analysis in Chapter 6. Experimental evidence is provided by a slight curvature of $\ln D$ versus $1/T$ curves for such data as copper self-diffusion. This slight curvature leads these researchers to fit the diffusivity to an equation of the form

$$D = D_{IV}\left(1 + D_{21} \exp -\frac{Q_{21}}{K_B T}\right)$$

where D_{IV} is the diffusion coefficient for single vacancies, and the term $D_{21} \exp(-Q_{21}/K_B T)$ is a correction for divacancies. In copper, for example, Mehrer and Seeger[3] find $\Delta H_m + \Delta H_v$, as contained in D_{IV}, to equal 2.09 e.v. and $Q_{21} = 0.51$ e.v. They have also found that D_{21} and Q_{21} are not negligible for the systems silver, gold, nickel, aluminum, and lead. Clearly, it appears that one must consider the influence of divacancies as the melting temperature is approached. At lower temperatures, the single vacancy mechanism is well-established in the close-packed metals like fcc.

ANOMALOUS DIFFUSION IN bcc CRYSTALS

Some essential features of diffusion in fcc metals that have been found experimentally may be contrasted with the unusual behavior in the bcc metals. These features are all consistent with diffusion by a single vacancy mechanism and are characterized by what is known to be "normal" diffusion. A plot of $\ln D$ versus $1/T$ yields a straight line through the diffusion data so that one may write

$$D = D_0 \exp -\frac{Q}{K_B T} \qquad\qquad 3.96$$

to fit fcc self-diffusion. Within this equation there are two features that appear to be consistent with all the fcc data. First, D_0 is within an order of magnitude of 1 cm^2/sec. Second, $Q \cong 34 T_m$ to within 20%, where T_m is the melting temperature.

For the "anomalous" behavior in the bcc metals, the plot of $\ln D$ versus $1/T$ is not linear for self-diffusion in the systems Z_r[4] and T_i.[5] Such a curvature is not consistent with Eq. 3.96 unless D_0 and Q are temperature dependent. A temperature dependence to these two quantities is not found in the fcc metals and, therefore, is not expected in the bcc metals.

As a consequence, some controversy exists for the diffusion in bcc metals. In the case of sodium,[6] the problem seems to be solved.

There are two sides to the controversy: those who believe that some new and complicated behavior is being seen in the bcc metals, and those who believe that a simple explanation exists, and is related to what is already known about diffusion in the fcc metals. This author is a member of the latter group, and therefore, presents that philosophy herein.

If one will not accept the temperature dependent D_0 and Q, then the sum of a number of exponentials shown in Eq. 3.97 is used to explain a curvature in the data plotted as ln D versus $1/T$.

$$D = \sum_i A_i \exp - \frac{Q_i}{K_B T} \qquad 3.97$$

In such an equation, two or three terms are usually sufficient. Each term represents a competing diffusion mechanism. One of the terms is probably due to the single vacancy mechanism so dominant in the fcc metals. There are several possibilities for the other terms, and different mechanisms may actually work in different systems.

For example, there is always a small population of interstitial atoms that are normally substitutional. Such an interstitial would diffuse by an interstitial mechanism very rapidly. This entire term would normally be negligible because the number of such interstitials is too low, and the A_i term for such a mechanism is proportional to the interstitial concentration. It is not suggested that the interstitial term causes the bcc anomaly; however, such a possibility cannot be easily excluded, as is discussed in Chapter 7. Therein some fcc metals are discussed, where interstitial diffusion of normal substitutional solutes is a practical necessity.

Table 3.3 presents the two- and three-exponential parameter fit to the diffusion in some of the bcc metals.

Table 3.3 Self-diffusion parameters in bcc metals

Metal	A_1 (cm²/sec)	Q_1 (kcal)	A_2 (cm²/sec)	Q_2 (kcal)	A_3 (cm²/sec)	Q_3 (kcal)
B–Zr[a]	1.34	65.2	8.5×10^{-5}	27.7		
B–Ti[b]	1.09	60.0	3.6×10^{-4}	31.2		
Na[c]	0.72	11.5	5.7×10^{-3}	8.53		
	8.0	14.0	1.4×10^{-1}	10.84	9×10^{-3}	8.72

[a] J. A. Federer, and T. S. Lundy, *ibid.*
[b] J. F. Murdock, T. S. Lundy, and E. E. Stansburg, *ibid.*
[c] J. N. Mundy, *ibid.*

It is notable that the numbers for A_1 and Q_1 in Table 3.3 are consistent with the criteria for "normal" diffusion, and these suggest that at least one of the mechanisms is that of a single vacancy. Additional evidence for the existence of vacancies participating in diffusion in bcc metals is the existence of a Kirkendall shift in these metals. The Kirkendall effect is discussed in the next chapter.

I feel that the major point of controversy is the mechanism causing the existence of the second or third exponential terms in the diffusivity. There is clear evidence that divacancies contribute to the self-diffusion of fcc silver at high temperatures. Divacancies, which are discussed in Chapter 5, may also contribute to the diffusion of bcc metals at high temperature. In fact, Peterson and Chen[7] recently discussed a model by Seeger that quantitatively explains the self-diffusion of sodium, a normal bcc metal, using a divacancy model. Seegers model requires three exponential terms. The first is for diffusion by single vacancies. The second and third are both divacancy tracer exchanges. A divacancy that consists of two neighboring vacancies can move by nearest neighbor jumps only if it partially dissociates to two vacancies at second neighbor sites on half of its jumps. This is entirely due to the nature of the bcc lattice, and yields two jump rates for the motion of the divacancy. The first is the vacancy jump from first to second neighbor sites and the second is the reverse. If there is any binding between the vacancies, these jumps occur at differing rates giving two exponential contributions to the diffusivity.

Another popular model for the existence of a second exponential term in the diffusivity is from Kidson.[8] He suggests that there could be a strong attraction between vacancies and some impurity in the metal. This would provide additional vacancies that are tied to the impurity, but could still exchange with the tracer. Thus the intrinsic vacancies would give one exponential contribution and the vacancy-impurity atom pairs could provide the second. He suggests that interstitial oxygen could be important in the instance of zirconium. This is an interesting idea because the experimental results and the crystal purity have appeared to vary between laboratories. Furthermore, the impurity must be relatively fast diffusing, like oxygen, since the vacancy would not be.

There are a number of other mechanisms that could explain the existence of a second exponential. The previous two appear to be the best ideas presented as yet, however. Due to the structure of the bcc lattices, one might consider vacancy interactions and jumps between second neighbor sites as a cause of the second exponential. This can possibly be excluded because the activation energy would necessarily be larger than the first neighbor exchange. Thus one may exclude such a model because

$A_2 \gg A_1$ and $Q_2 > Q_1$ in the table. However, one can not entirely disregard such a model; it has been proposed to explain some bcc data based on second neighbor vacancy solute binding and its influence on solvent diffusion. The presence of the solute will alter the diffusion of the solvent (see Chapter 5).

As mentioned earlier, one can not totally ignore the possible existence of interstitial solutes, and this gives three possible mechanisms to explain the anomalous bcc data. Some very clever experimental work is evidently needed in the bcc metals.

REFERENCES

1. J. R. Manning, *Phys. Rev.* **124,** 470 (1961).
2. J. P. Stark, *Acta Met.* **22,** 533 (1974).
3. H. Mehrer and A. Seeger, *Phys. Stat. Solids* **35,** 313 (1969).
4. J. A. Federer and T. S. Lundy, *Trans. AIME,* **227,** 2341, (1963) and G. V. Kidson, *Can. J. Phys.* **41,** 1553 (1963).
5. J. F. Murdock, T. S. Lundy, and E. E. Stansburg, *Acta Met.,* **12,** 1033, (1964).
6. J. N. Mundy, *Phys. Rev.,* **B3,** 2431 (1971).
7. N. L. Peterson and W. K. Chen, *Ann. Rev. Mat. Sci.* **3,** 75 (1973).
8. G. V. Kidson, *Can. J. Phys.* **41,** 1563 (1963).

CHAPTER 4

THE INFLUENCE OF INTERNAL AND EXTERNAL FIELDS UPON DIFFUSION BY A VACANCY MECHANISM

PHENOMENOLOGICAL EQUATIONS

Perhaps the easiest way to understand the influence of an applied field is through the phenomenological equations for mass and energy transport, as introduced in Chapter 2. Thus one starts with a minor generalization of Eq. 2.95, where the generalized forces cause mass transport, and self-diffusion is introduced as a limiting case somewhat after the fact. The generalization is the introduction of an external field; whereas only internal fields are described in Eq. 2.95. These are the isothermal chemical potential gradients and the temperature gradient.

The fluxes for an n component substitutional alloy may be written as follows:

$$J_k = \sum_{j=1}^{n} L_{kj}x_j + L_{kv}x_v + L_{kq}x_q$$

$$J_v = \sum_{j=1}^{n} L_{vj}x_j + L_{vv}x_v + L_{vq}x_q$$

$$J_q = \sum_{j=1}^{n} L_{qj}x_j + L_{qv}x_v + L_{qq}x_q \qquad 4.1$$

The X_j are the generalized forces, the subscript v denotes the vacancies, and the subscript q denotes the heat flow. The generalized forces may be written in terms of the isothermal chemical potential gradients, $\nabla \mu_j / T$

$$x_q = -\frac{1}{T} \nabla T$$

$$x_v = -\nabla \mu_v / T$$

and

$$x_j = -\nabla \mu_j / T + F_j \qquad 4.2$$

where F_j is the influence of an external field. As an example of one such external field, consider an electric potential gradient. In that particular instance, the external force on the diffusing ion is the following:

$$F_j = q_j E \qquad 4.3$$

where q_j is the effective ionic charge of an atom of species j, and E is the electric field intensity.

It is important to remember that all the above forces are independent quantities since one is viewing the return of a system to equilibrium from some fluctuation. The introduction of microscopic reversibility so that the Onsager reciprocal relations may be used is assumed relative to the forces since they exist as independent parameters. As a consequence, microscopic reversibility has no influence on the forces themselves. This important point is easily understood for all the atomic species present; however, the vacancies present another problem which is discussed later.

Microscopic reversibility is therefore assumed to exist in the system, with the above limitation in mind. There are two restrictions that one can consequently use to reduce the above equations. We assume an alloy of substitutional atoms comprises the system. In any region of the crystal where vacancies are neither being generated or annihilated, the sum of the fluxes must add to zero. Thus

$$J_v = -\sum_{k=1}^{n} J_k \qquad 4.4$$

Substituting Eq. 4.1 into Eq. 4.4, one concludes that

$$L_{vj} = -\sum_{k=1}^{n} L_{kj}$$

$$L_{vv} = -\sum_{k=1}^{n} L_{kv}$$

$$L_{vq} = -\sum_{k=1}^{n} L_{kq} \qquad 4.5$$

If one uses the second restriction that the Onsager reciprocal relations are valid, then

$$L_{vj} = L_{jv}, \; L_{kj} = L_{jk}$$
$$L_{vq} = L_{qv},$$

and

$$L_{kq} = L_{qk}. \qquad 4.6$$

Using Eq. 4.6 in Eq. 4.5,

$$L_{kv} = -\sum_{j=1}^{n} L_{kj}$$

$$L_{qv} = -\sum_{j=1}^{n} L_{qj} \qquad 4.7$$

Equations 4.7 may be used to reduce Eq. 4.1 to the following

$$J_k = \sum_{j=1}^{n} L_{kj}(x_j - x_v) + L_{kq}x_q$$

$$J_q = \sum_{j=1}^{n} L_{qj}(x_j - x_v) + L_{qq}x_q \qquad 4.8$$

The diffusive jump of each atom carries some heat with it that can be measured through a properly designed experiment. Letting this magnitude of energy per unit diffusive jump of component i be Q_i, one can define

$$L_{kq} = \sum_{j=1}^{n} Q_j L_{kj} \qquad 4.9$$

Equation 4.9 can be used to simplify Eq. 4.8 to the following:

$$J_k = \sum_{j=1}^{n} L_{kj}(x_j - x_v + Q_j x_q) \qquad 4.10a$$

$$J_q = \sum_{i=1}^{n} Q_i J_i + \left(L_{qq} - \sum_{j=1}^{n} L_{jq}Q_j\right)x_q \qquad 4.10b$$

It is apparent from Eq. 4.10 that the energy Q_i is the heat carried for a diffusive jump. This can be seen by letting x_q equal zero, in which case only the diffusive motion of the atoms can carry the heat, as is properly predicted by Eq. 4.10b. Equation 4.10a gives the atom flux for a generalized force; further reduction of this equation necessitates a specific example. Consequently, we consider a binary alloy, which is the easiest situation to deal with rigorously.

DEFINITION OF SELF-DIFFUSION

A binary alloy is the mixture of isotopes of two chemical species, A and B. Self-diffusion refers to the motion of an isotope tracer of one element, say B, through an alloy that is homogeneous in the concentration of A in the absence of any applied external field. Thus to identify the important parameters in self-diffusion, one must discuss a pseudoternary alloy made up of component A, component B, and a very dilute tracer of B, denoted B^* and then follow the flux of the tracer. Since there are no external forces and the system is assumed to be at uniform temperature and composition of A, one can deduce the chemical potential gradients as are implied within Eq. 4.10. That is, since the temperature is uniform, $x_q = 0$. There exists no field or force to perturb the concentration of vacancies from a uniform distribution; this implies that $x_v = 0$. The other chemical potentials, μ_i, $i = A$, B, and B^* can be deduced from

$$\mu_i = \mu_{i0} + k_B T \ln C_i \gamma_i \qquad 4.11$$

where μ_{i0} is a standard state. Thus

$$\nabla \mu_i = k_B T \left(\frac{\nabla C_i}{C_i} + \nabla \ln \gamma_i \right) \qquad 4.12$$

For the homogeneous component A, $\nabla C_A = 0$ by definition. This also implies that $\nabla \ln \gamma_A = 0$, so that $\nabla \mu_A = 0$. The activity coefficients of B and B^* are sufficiently identical that one can interchange an atom of B and B^* without seriously perturbing the local free energy of the crystal. The vibrational energy and entropy of B and B^* differ slightly; however, this effect is ignored. As a consequence, $\nabla \ln \gamma_B = \nabla \ln \gamma_{B^*} = 0$. Thus $\nabla \mu_A = 0$ and the equations

$$\nabla \mu_B = K_B T \frac{\nabla C_B}{C_B}$$

$$\nabla \mu_{B^*} = K_B T \frac{\nabla C_{B^*}}{C_{B^*}} \qquad 4.13$$

and

$$\nabla C_B = -\nabla C_{B^*} \qquad 4.14$$

define the conditions for self-diffusion of tracer B^*. Hence Eq. 4.10a reduces to

$$J_{B^*} = -L_{B^*B^*} \nabla \mu_{B^*} - L_{B^*B} \nabla \mu_B$$

$$= \frac{-K_B T L_{B^*B^*}}{C_{B^*}} \left(1 - \frac{L_{B^*B} C_{B^*}}{L_{B^*B^*} C_B} \right) \nabla C_{B^*} \qquad 4.15$$

Also, by definition of the self-diffusion coefficient, under identical circumstances one has

$$J_{B^*} = -D_{B^*} \nabla C_{B^*} \qquad 4.16$$

These two equations may be compared to imply

$$D_{B^*} = \frac{K_B T L_{B^*B^*}}{C_{B^*}} \left(1 - \frac{L_{B^*B} C_{B^*}}{L_{B^*B^*} C_B} \right) \qquad 4.17a$$

$$\cong \frac{K_B T L_{B^*B^*}}{C_{B^*}} \qquad 4.17b$$

where the latter approximation follows since $C_{B^*} \ll C_B$ by hypothesis. Equation 4.17 therefore defines the self-diffusion coefficient in terms of the phenomenological coefficient.

INTRINSIC DIFFUSION OF A TRACER

When a tracer atom moves relative to the lattice planes in a chemical potential gradient with no externally applied field and with a uniform temperature, intrinsic diffusion is said to occur. We therefore wish to identify what is meant by the intrinsic diffusion coefficient, and this can be done using Eq. 4.10. Since we are considering a pseudoternary as in the last section, it is important to first identify the quantities that restrict the tracer, B^*, to move at the same rate as B in the A, B, and B^* alloy. Again B^* is a very dilute component. Actually B^* will not move identically at the same rate as B due to differences in isotopic mass; the jump frequencies also differ slightly. We construct here a hypothetical situation in which the restrictions on the pseudoternary make it behave as a binary alloy.

For B^* to behave as B, it must respond to the same internal force as that of B. Since one of the forces tending to move B^* is the activity coefficient gradient, it must be the same for B and B^*. Thus we set

$$\frac{d \ln \gamma_{B^*}}{dx} = \frac{d \ln \gamma_B}{dx} \qquad 4.18$$

This statement must be expanded to include the concentration gradients because we ultimately wish to make a statement similar to Fick's first law. Therefore,

$$\frac{d \ln \gamma_{B^*}}{d \ln C_{B^*}} \left(\frac{1}{C_{B^*}} \frac{dC_{B^*}}{dx} \right) = \frac{d \ln \gamma_B}{d \ln C_B} \left(\frac{1}{C_B} \frac{dC_B}{dx} \right) \qquad 4.19$$

If the concentration of B^* is everywhere proportional to the concentration of B, and if we hypothesize that it remains so as time proceeds throughout an experiment, then no isotope separation occurs. Furthermore, establishing a proportionality such as

$$C_B = E_0 C_{B^*} \qquad 4.20$$

insures that

$$d \ln C_{B^*} = d \ln C_B$$

so that the thermodynamic factors

$$\frac{d \ln \gamma_{B^*}}{d \ln C_{B^*}} = \frac{d \ln \gamma_B}{d \ln C_B} = \alpha - 1 \qquad 4.21$$

Then one may treat the alloy as a binary, and the thermodynamic factor α may be determined for component A from the Gibbs'-Duhem equation as

$$\alpha = \frac{d \ln \gamma_A}{d \ln C_A} + 1 \qquad 4.22$$

All of the above is contingent on the validity of Eq. 4.20. It implies that

$$C_B = E_0 C_{B^*}$$

$$J_B = E_0 J_{B^*} \qquad 4.23a$$

$$\nabla \ln C_B = \nabla \ln C_{B^*} \qquad 4.23b$$

$$\nabla C_B = E_0 \nabla C_{B^*} \qquad 4.23c$$

Furthermore,

$$\nabla C_A = -\nabla(C_B + C_{B^*} + C_v)$$

$$\cong -\nabla(C_B + C_{B^*}) \qquad 4.24$$

where C_v is the negligible vacancy concentration. If the vacancies are in local equilibrium throughout the sample, the isothermal chemical potential gradient of the vacancies is zero. We assume this is sufficiently true that such an approximation will not alter our results. Hence

$$x_v = 0 \qquad 4.25$$

The flux of the tracer B^* is therefore

$$J_{B^*} = -L_{B^*B^*}\nabla\mu_{B^*} - L_{B^*B}\nabla\mu_B - L_{B^*A}\nabla\mu_A$$

$$= -L_{B^*B^*}K_BT\alpha\left\{\frac{\nabla C_{B^*}}{C_{B^*}} + \frac{L_{B^*B}}{L_{B^*B^*}}\frac{\nabla C_B}{C_B} + \frac{L_{B^*A}}{L_{B^*B^*}}\frac{\nabla C_A}{C_A}\right\} \qquad 4.26$$

where Eqs. 4.11, 4.21 and 4.22 have been used. On inserting Eq. 4.23b,

$$J_{B^*} = \frac{-L_{B^*B^*}K_BT\alpha}{C_{B^*}}\left\{1 + \frac{L_{B^*B}}{L_{B^*B^*}} + \frac{L_{B^*A}C_{B^*}\cdot\nabla C_A}{L_{B^*B^*}C_A\nabla C_{B^*}}\right\}\nabla C_{B^*} \qquad 4.27$$

Since Eqs. 4.23b, 4.23c, and 4.24 are valid, one can show that

$$\frac{\nabla C_A}{\nabla C_{B^*}} = -(1 + E_0)$$

and

$$\frac{C_{B^*}}{C_A} = \frac{C/C_A - 1}{1 + E_0} \qquad 4.28$$

where the total concentration, C, is $C = C_A + C_B + C_{B^*}$. Equation 4.28 implies that

$$\frac{C_{B^*}}{C_A} \frac{\nabla C_A}{\nabla C_{B^*}} = -\frac{N_B}{N_A} \qquad 4.29$$

where N_A and N_B are the mole fractions of components A and B, respectively. This term finally reduces Eq. 4.27 to

$$J_{B^*} = \frac{-L_{B^*B^*} K_B T \alpha}{C_{B^*}} \left\{ 1 + \frac{L_{B^*B}}{L_{B^*B^*}} - \frac{N_B}{N_A} \frac{L_{B^*A}}{L_{B^*B^*}} \right\} \nabla C_{B^*} \qquad 4.30$$

$$= -D^I_{B^*} \nabla C_{B^*}$$

The latter expression defines the intrinsic diffusion coefficient of the tracer B^* as

$$D^I_{B^*} = D_{B^*} \alpha r_{B^*} \qquad 4.31$$

where

$$r_{B^*} = 1 + \frac{L_{B^*B}}{L_{B^*B^*}} - \frac{N_B}{N_A} \frac{L_{B^*A}}{L_{B^*B^*}} \qquad 4.32$$

when one uses Eq. 4.17b for the self-diffusion coefficient of B^*. The latter factor, r_{B^*} is called the vacancy wind factor, as becomes apparent from further discussion latter in the chapter. In Eq. 4.31, the thermodynamic factor, α, is different from unity in concentrated systems. Also, r_{B^*} differs from unity under most situations. Thus the diffusion of B^* in a chemical potential gradient, as reflected by $D^I_{B^*}$, differs from self-diffusion in two ways. First, the thermodynamic factor reflects the fact that B^* is responding to a force involving the activity coefficient gradients. Second, the rate of motion of A and B will differ, as suggested by the self-diffusivities of a solvent and dilute solute in the last chapter. If the solvent and solute move at different rates, a vacancy wind is established, which influences the mobility of B^*. Thus B^* will tend to drift in one direction or the other because the vacancies are approaching B^* from one direction more frequently. In a later section, explicit expressions for these terms are derived for the dilute alloy treated in Chapter 3.

For a random concentrated alloy for which there is no solute-vacancy binding, Manning[1] has analyzed the vacancy wind factors and finds that

$$D_i^I = D_i^* \alpha r_i \qquad 4.33$$

where

$$r_A = \frac{1 + 2N_A(D_A^* - D_B^*)}{M_0(N_A D_A^* + N_B D_B^*)} \qquad 4.34$$

and r_B is the same expression when A and B are interchanged. Here M_0 is a pure number that depends on crystal structure having the value 7.15 for fcc, 5.33 in bcc, and so forth. It is clear that the vacancy wind factor, r_i, depends on the physical effect that the isotopes of A and B differ.

KIRKENDALL EFFECT AND MUTUAL DIFFUSIVITY

When the self-diffusion coefficients of components A and B differ, which is almost always the case, the lattice planes will be moving relative to the end of the specimen at rest. The intrinsic diffusion coefficients are related to this coordinate system moving at a velocity V_L, as mentioned in Chapter 1. The motion of these planes is due to the net drift of vacancies from one end of the sample to the other. The flux of A or B relative to the fixed laboratory frame of reference is needed in the conservation of mass Eq. 1.8. Thus one measures diffusion coefficients relative to the fixed end of the specimen, which differ from those moving relative to the lattice planes. The shift of the lattice planes is needed to determine the intrinsic diffusivities. For this purpose, inert markers are welded into the sample during specimen preparation. Then when the lattice planes shift position relative to the end of the sample, one can measure the net distance moved.

To analyze this problem, one needs the flux of the two components relative to the fixed end of the sample. The lattice planes move at a velocity V_L; hence

$$J_i' = -D_i^I \frac{\partial C_i}{\partial x} + C_i V_L \qquad 4.35$$

The total atom flux, J_t, which is the sum of the fluxes of both components, becomes

$$J_t = J_A' + J_B' = (D_B^I - D_A^I) \frac{\partial C_A}{\partial x} + (C_A + C_B)V_L \qquad 4.36$$

One may also define the flux relative to the fixed ends of the sample as

$$J_i' = -D_i{}^m \frac{\partial C_i}{\partial x} \qquad 4.37$$

so that

$$J_t = (D_B{}^m - D_A{}^m)\frac{\partial C_A}{\partial x} \qquad 4.38$$

This flux, J_t, is zero by definition when the volume change on mixing is zero. We assume that this is so; thereby from Eqs. 4.35–4.38

$$V_L = (D_A{}^I - D_B{}^I)\frac{\partial N_A}{\partial x} \qquad 4.39$$

and define the mutual diffusivity, $\tilde{D}$, as

$$\tilde{D} = D_A{}^m = D_B{}^m = N_B D_A{}^I + N_A D_B{}^I \qquad 4.40$$

If one substitutes Eqs. 4.33 and 4.34 into Eq. 4.40, as was done by Manning, then

$$\tilde{D} = (N_B D_A^* + N_A D_B^*)\alpha R \qquad 4.41a$$

where

$$R = 1 + \frac{\alpha N_A N_B (D_A^* - D_B^*)}{M_0(N_A D_B^* + N_B D_A^*)(N_A D_A^* + N_B D_B^*)} \qquad 4.41b$$

These vacancy wind factors and phenomenological coefficients can be found for dilute alloys using a kinetic theory for the mobility of a substitutional tracer, as discussed in the next section.

KINETIC THEORY FOR MOBILITY[2]

To follow the motion of a tracer atom in an applied field and include the vacancy wind effect, one must start at the source of the defect and let it approach the tracer. The evolution of the fluctuation continues until the last tracer defect exchange has occurred.

To this end, assume that a single defect has originated at a source. It will move up to the tracer, cause a sequence of correlated tracer exchanges, and migrate to its sink. The tracer will move relative to a field applied in the $+x$ direction. The crystal containing the tracer will be face-centered cubic, and the x axis will be perpendicular to (100) planes. With a unit cell dimension of $2b$, any single defect-tracer exchange with nonzero x projection will result in a tracer motion of $\pm b$ relative to the x axis.

The formalism developed is capable of dealing with more complex problems than the vacancy-solute pair mentioned in the last chapter. The notation is a matrix generalization of the terms P_{11} and P_a used therein. The generalization counts other possible configurations that could arise from such possibilities as divacancies altering the diffusivity and mobility; in addition, the applied field influences the probabilities P_{11} and P_a.

The defect sources and sinks will be uniformly distributed at some large distance from the tracer. The tracer is at the center of a sphere whose surface contains defect sources and sinks.

In general, there will be a number of tracer-defect configurations that may lead to a single tracer jump with nonzero x projection. These will be denoted as α, $\alpha = 1, \cdots N$; for example, in fcc there is one configuration for a single solute-vacancy pair, two configurations for a bound divacancy, and thirteen configurations for a vacancy pair that may dissociate at the solute. Figure 4.1 illustrates the bound divacancy jump configurations. Immediately following a tracer defect jump that starts from configuration $-\alpha$, the tracer and defect will be in a configuration β. From this configuration, a second tracer jump can occur directly, a β-type jump. On the other hand, the defect could wander off and return to the tracer forming a configuration, γ, and cause the tracer jump.

Therefore, one assumes the defect starts its motion at the source and migrates up to the tracer, forming a configuration that can lead to a tracer-defect exchange with nonzero x projection. The configuration so formed can cause a direct tracer exchange in the $+x$ direction. Since the tracer is in a jump configuration, it is also in the particular configuration that would follow a jump of type $-\alpha$ in the direction $-x$. Consequently, one defines the rate of formation of the complex immediately following a

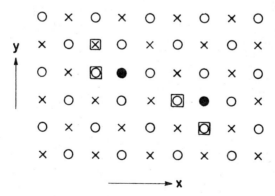

Figure 4.1 Bound divacancy jump configurations in fcc are denoted as α ($\alpha = 1, 2$). Mirror image configurations are α^I ($\alpha^I = 1, 2$). ($\square$ vacancy, $\bullet$ solute) (100) plane pictured.

tracer defect jump from a $-\alpha$ configuration as $\nu_{\pi-\alpha}$. The tracer is thus in a position to make a jump from a configuration $+\beta$ in the positive x direction.

At this point, several things could happen, and one must define additional terms. Because the field is in the $+x$ direction, the tracer could exchange with the defect in the $+x$ direction parallel to the field but antiparallel to the hypothetical first jump of $-\alpha$. Consequently, $P_{\alpha\beta}^{+a}$ is defined as the effective probability of a β type jump in the "+" field direction and antiparallel, "a," to the previous jump of type α. Instead of the previous jump, the defect could leave the tracer and wander around to the other side without exchanging with the tracer, and cause a β jump parallel with the original α jump but in the "$-$" field direction. One, therefore, defines $P_{\alpha\beta}^{-11}$ as the effective probability that following an α tracer jump in the minus field direction, the next tracer defect exchange will be in the $-$ field direction but parallel, "11," with the last tracer jump. For a sequence of tracer defect exchanges one must similarly define $P_{\alpha\beta}^{-a}$ and $P_{\alpha\beta}^{+11}$ as follows: $P_{\alpha\beta}^{-a}$ is the effective probability that a β jump will occur that is antiparallel to the previous α jump, but in the negative field direction; the α jump would have been in the $-$ field direction. We define $P_{\alpha\beta}^{+11}$ in a similar manner.

If the defect first arrives at the tracer in a configuration that follows a jump of $-\alpha$, without previously exchanging with the tracer and without arriving at any other configuration that could cause a tracer jump in the x direction, then only two possible tracer jumps could occur: (1) the defect could yield a tracer displacement of $+b$ with an effective probability $P_{\alpha\beta}^{+a}$; (2) the tracer could move a distance $-b$ with a probability of $P_{\alpha\beta}^{-11}$.

Assuming that (1) occurs, the second tracer jump could: (3) give a net displacement of $+2b$ with a probability $P_{\alpha\beta}^{+11}$, or (4) a net displacement of $0b$ with a probability $P_{\beta\gamma}^{-a}$. On the other hand, if (2) occurs the second tracer jump could: (5) make a net displacement of $-2b$ with a probability $P_{\beta\gamma}^{-11}$, or (6) $0b$ with a probability $P_{\beta\gamma}^{+a}$.

The probability that (1) occurs and is followed by the defect migrating to a sink is evidently given by

$$P_{\alpha\beta}^{+a}\left(1 - \sum_{\gamma} P_{\beta\gamma}^{+11} - \sum_{\gamma} P_{\beta\gamma}^{-a}\right) \qquad 4.42$$

The term in parenthesis is the probability that no further jumps occur, and it implies that the β jump was the first and last tracer jump.

With N jump configurations, one arranges the elements $P_{\alpha\beta}^{+a}$ into a matrix P^{+a}, $N \times N$. Let 1 be the unit column vector; then Eq. 4.42 may be written as

$$P^{+a}(I - P^{+11} - P^{-a})1 \qquad 4.43$$

I is the N-dimensional unit matrix.

On the other hand, process (2) could occur and be followed by the defect going to a sink. The effective probability for this is

$$P^{-11}(I - P^{-11} - P^{+a})1 \qquad\qquad 4.44$$

The average x displacement for the first tracer-defect exchange, assuming it is the last exchange, is

$$b\{P^{+a}(I - P^{+11} - P^{-a}) - P^{-11}(I - P^{-11} - P^{+a})\}1 \qquad 4.45$$

Dividing Eq. 4.45 by b gives the net probability that the tracer will make a displacement of $+b$ on the first tracer jump, and that the defect will then go to a sink.

For an extended sequence of tracer-vacancy exchanges, an equation similar to Eq. 4.45 divided by b may be interpreted as the net probability that the tracer makes an average x displacement of $+b$. This matrix is defined as $S+$ in the following:

$$
\begin{aligned}
S+ =\ & P^{+a}(I - P^{+11} - .P^{-a}) - P^{-11}(I - P^{-11} - P^{+a}) \\
& + 2P^{+a}P^{+11}(I - P^{+11} - P^{-a}) + 0P^{+a}P^{-a}(I - P^{-11} - P^{+a}) \\
& + 0P^{-11}P^{+a}(I - P^{+11} - P^{-a}) - 2P^{-11}P^{-11}(I - P^{-11} - P^{+a}) \\
& + 3P^{+a}P^{+11}P^{+11}(I - P^{+11} - P^{-a}) + P^{+a}P^{+11}P^{-a}(I - P^{-11} - P^{+a}) \\
& - P^{+a}P^{-a}P^{-11}(I - P^{-11} - P^{+a}) + P^{+a}P^{-a}P^{+a}(I - P^{+11} - P^{-a}) \\
& + P^{-11}P^{+a}P^{+11}(I - P^{+11} - P^{-a}) - P^{-11}P^{+a}P^{-a}(I - P^{-11} - P^{+a}) \\
& - P^{-11}P^{-11}P^{+a}(I - P^{+11} - P^{-a}) - 3P^{-11}P^{-11}P^{-11}(I - P^{-11} - P^{+a}) \\
& + \cdots \qquad\qquad\qquad\qquad\qquad\qquad\qquad\qquad\qquad\qquad 4.46
\end{aligned}
$$

Many of the terms in Eq. 4.46 cancel, and one can write it as

$$
\begin{aligned}
S+ =\ & P^{+a}\{I + P^{+11} + P^{+11}P^{+11} + P^{-a}P^{+a} + P^{+11}P^{+11}P^{+11} \\
& + P^{+11}P^{-a}P^{+a} + P^{-a}P^{-11}P^{+a} + P^{-a}P^{+a}P^{+11} + \cdots\} \\
& + P^{-11}\{P^{+a} + P^{-11}P^{+a} + P^{+a}P^{+11} + P^{-11}P^{-11}P^{+a} \\
& + P^{-11}P^{+a}P^{+11} + P^{+a}P^{+11}P^{+11} + P^{+a}P^{-a}P^{+a} + \cdots\} \\
& - P^{-11}\{I + P^{-11} + P^{-11}P^{-11} + P^{+a}P^{-a} + P^{-11}P^{-11}P^{-11} \\
& + P^{-11}P^{+a}P^{-a} + P^{-a}P^{+11}P^{-a} + P^{+a}P^{-a}P^{+11} + \cdots\} \\
& - P^{+a}\{P^{-a} + P^{+11}P^{-a} + P^{-a}P^{-11} + P^{+11}P^{+11}P^{-a} + \\
& + P^{+11}P^{-a}P^{-11} + P^{-a}P^{-11}P^{-11} + P^{-a}P^{+a}P^{-a} + \cdots\} \\
=\ & P^{+a}\{F_{(+)}\} + P^{-11}\{G_{(+)}\} - P^{-11}\{F_{(-)}\} - P^{+a}\{G_{(-)}\} \\
=\ & P^{+a}\{F_{(+)} - G_{(-)}\} - P^{-11}\{F_{(-)} - G_{(+)}\} \qquad\qquad 4.47
\end{aligned}
$$

A closed form for the four matrix series in Eq. 4.47 is not possible with the dimensionality of the matrices present; one must go to a $2N \times 2N$ matrix. One can follow the evolution using a random walk of the defect in a manner similar to the approach taken previously for P_{11} and P_a in the calculation of D.

The total evolution of the fluctuation can be followed by first writing the rate of formation of the complex following a $-\alpha$ jump, $\nu_{\pi-\alpha'}$ as a diagonal matrix $\nu(-)$ with elements $\nu_{\pi-\alpha}$. Now the effective jump frequency of the tracer in the $+x$ direction, a scalar, ν_{+e}, is found from the following:

$$\nu_{+e} = 1^T \nu_{(-)} S_+ 1$$

$$= \sum_{\beta=1}^{N} \sum_{\alpha=1}^{N} \nu_{\pi-\alpha}(S_+)_{\alpha\beta} \qquad\qquad 4.48$$

where 1^T is the transpose of 1.

Prior to investigating the terms in Eq. 4.48, it is of interest to examine Eq. 4.47 in a limiting case.

The limiting case is the situation where the applied field is removed. Then, $P^{+11} = P^{-11}$ and $P^{+a} = P^{-a}$. Consequently, using a subscript 0 to denote the absence of the field,

$$S_{+0} = (P^a - P^{11})[1 - (P^a - P^{11}) - (P^a - P^{11})^2 - (P^a - P^{11})^3 + \cdots]$$

$$S_{+0} = (P^a - P^{11})[I + (P^a - P^{11})]^{-1}$$

Letting $P = P^a - P^{11}$ yield

$$S_{+0} = P[I + P]^{-1} = P(I - P)^{-1}\{(I - P)(I + P^{-1}\}$$

$$= P(I - P)^{-1}f$$

$$f = (I - P)(I + P)^{-1} = I - 2S_{+0} \qquad\qquad 4.49$$

where f is the generalized matrix formulation for the correlation factor. The matrix form is discussed in Chapter 5. For comparison, Eqs. 3.45 and 4.49 are identical when $N = 1$.

Linearization of the Effective Jump Frequency

The influence of an applied field on the motion of a tracer is expected to be proportional to the force. Equation 4.48 contains nonlinear terms resulting from the product of probability terms, such as those in Eq. 4.47. In the expansion of the field effect it is therefore important to delete all nonlinear terms in the force. For that purpose, Eq. 4.48 can be written as

$$\nu_{+e} = 1^T\{(\nu_{(-)} - \nu_{(-)0}) + \nu_{(-)0}\}\{(S_+ - S_{+0}) + S_{+0}\}1$$

where the subscript 0 implies the absence of the field. If one expands the terms in braces to exclude all nonlinear field effects, then to the first order in the applied force one finds

$$\nu_{+e} = 1^T\{\nu_{(-)0}S_{+0} + (\nu_{(-)} - \nu_{(-)0})S_{+0} + \nu_{(-)0}(S_+ - S_{+0})\}1$$
$$= 1^T\{I + \delta\nu_{(-)}\nu_{(-)0}^{-1} + \nu_{(-)0}\,\delta S_+ S_{+0}^{-1}\nu_{(-)0}^{-1}\}\nu_{(-)0}S_{+0}1 \qquad 4.50$$

In Eq. 4.50, $\delta\nu$ and δS are proportional to the force to the first power. Also in Eq. 4.50, any terms associated with a diffusion coefficient gradient are omitted. Chapter 2 showed that the diffusion coefficient gradient does not contribute to the tracer flux, and this argument is independent of diffusion mechanism. Since these terms are omitted in Eq. 4.50, the drift velocity of the tracer resulting from the force is $b(\nu_{+e} - \nu_{-e})$. Since the field effect terms in Eq. 4.50 are linear in the force, $\delta\nu_+ = -\delta\nu_-$ and $\delta S_+ = -\delta S_-$. Consequently,

$$V_F = b(\nu_{+e} - \nu_{-e}) = b1^T(2\delta\nu_{(-)}S_0 + 2\nu_{0(-)}\delta S_+)1$$
$$= b1^T(2\delta\nu_{(-)}\nu_{(-)0}^{-1} + \nu_{(-)0}(\delta S_+ - \delta S_-)S_0^{-1}\nu_{(-)0}^{-1})\nu_{(-)0}S_01 \qquad 4.51$$

Hence the tracer flux is found from

$$J_K = -D_K^* \frac{\partial c_K}{\partial x} + c_K V_F \qquad 4.52$$

where c_K is the concentration.

To evaluate Eqs. 4.51 and 4.52 and, thereby, the effective charge for electromigration or the heat of transport for thermomigration, a set of defect walk matrices similar to the previous random walk matrices used in Chapter 3 must be defined. For such a matrix one must identify a set of defect sites surrounding the tracer, which includes a region of the crystal. Then if there are M such sets of configurations, $R_n(i)$ is the probability that the defect configuration i is occupied after the nth defect jump from some arbitrary origin. Here $R_n(i)$ is an M-component column vector; M is the number of defect configurations. A matrix A is defined in which the matrix elements, A_{ij}, are the probability that the *particular* defect site in class i is occupied after n jumps, provided unit occupation probabilities for sites in class j are given after $n-1$ jumps. Then

$$R_n = AR_{n-1} \qquad 4.53$$

If R_0 is the initial probability distribution, and $W_{(\alpha)}$ is the sum of the probabilities over all n written as a column vector

$$W_{(\alpha)} = \sum_{n=0}^{\infty} R_n = (I - A)^{-1}R_0 \qquad 4.54$$

The matrix A and others like it will be dependent on the applied field. One must, therefore, be able to reduce an equation such as 4.54 to one that is linear in the field. For this purpose, it is clear that the matrix A can be written as the sum of two terms; $A = A_0 + \delta A$, where A_0 is independent of the applied field and δA is proportional to the field strength. From Eq. 4.54 one can easily show that

$$W_{(\alpha)} = (I - A_0)^{-1} R_0 + (I - A_0)^{-1} \delta A (I - A_0)^{-1} R_0 \qquad 4.55$$

The first term in Eq. 4.55 is independent of the field strength; the second term is proportional to it. Higher-order terms in $(1 - A)^{-1}$, such as $(\delta A)^2$, are neglected in such an expansion.

The first transition probability matrix of interest will be A_2, representing M-defect configurations. The A_2 is essentially the same matrix defined by Howard[3] except for two important differences. First, A_2 includes the tracer jumps and second, A_2 is linear in the applied field as in Eq. 4.55. Since tracer jumps are included in A_2 it is important to write the matrix for classes of defect sites. When this is done, a tracer jump will move the set of defect configurations to a new origin each time a tracer jump in the x direction occurs.

The matrix A_2 can be used to evaluate S_+ in Eq. 4.47 and consequently δS_+ and S_0 in Eq. 4.51. In Eq. 4.47 the first term $P^{+a} F_{(+)}$ is evidently the sum of the probabilities that a sequence of tracer-defect exchanges ends with a + tracer jump, with respect to the direction of the field, provided the sequence starts with a + tracer jump from a configuration that normally gives a + tracer jump on the first exchange. Here $P^{-11} G_{(+)}$ is the sum of the probabilities that a sequence of tracer-defect exchanges starts with the first tracer exchange in the − field direction, with a defect whose path originated from an orientation that normally leads to a + tracer exchange on the first jump, and the sequence ends with a + tracer exchange. The terms $P^{+a} G_{(-)}$ and $P^{-11} F_{(-)}$ are defined in a similar manner. In addition one defines $R_{(\alpha)}$ as a column vector of occupation probabilities following a tracer jump of type α; $R_{(\alpha)}$ is the tracer jump from the image configuration. Furthermore, $Q(\beta)$ is the tracer transition probability for a β type jump written as a row vector; $Q(\beta^I)$ is the reverse jump, and both of these later terms can be written as $Q_0(\beta) + \delta Q(\beta)$, where δQ is linear in the applied field. With the above definition, one can follow a set of tracer-defect exchanges and determine the terms in Eq. 4.47. The matrix elements of the terms in S_+ are found in the following:

$$(P^{+a} F_{(+)} + P^{-11} G_{(+)})_{\alpha\beta} = Q(\beta)(I - A_2)^{-1} R(\alpha^I)$$
$$(P^{-a} F_{(-)} + P^{+11} G_{(-)})_{\alpha\beta} = Q(\beta^I)(I - A_2)^{-1} R(\alpha)$$
$$(P^{+a} G_{(-)} + P^{-11} F_{(-)})_{\alpha\beta} = Q(\beta^I)(I - A_2)^{-1} R(\alpha^I)$$
$$(P^{-a} G_{(+)} + P^{+11} F_{(+)})_{\alpha\beta} = Q(\beta)(I - A_2)^{-1} R(\alpha) \qquad 4.56$$

From prior discussion, the right hand side of Eq. 4.56 can be linearized with respect to the applied force. Equations 4.56 provide the necessary information to evaluate S_0 and δS.

To evaluate $\nu_{(-)}$, and hence $\delta \nu_{(-)}$ and $\nu_{(-)0}$, go to the set of defect sites one defect jump beyond the limits of the class of sites incorporated in A_2. These defect sites outside A_2 are assumed to be the source sites. Then, define $T(\gamma)$ as the number of defects per unit time jumping from a set of outer defect sites into a particular defect site, γ, contained in the set of sites included in A_2. Thus $T(\gamma)$ is linear in the field, and it may be written as a column vector. Now $\nu_{\pi-\alpha}$ is the rate that defects leave the outer defect sites and arrive at configuration $-\alpha$ without exchanging with the tracer, and without arriving at any other site where a tracer exchange can occur. Hence using $T(\gamma)$ one can find the probability that the defect path will reach $-\alpha$; the probability that the defect goes from γ to $-\alpha$ is $H_{-\alpha\gamma}$. Thus one expects

$$\nu_{\pi-\alpha} = \sum_{\gamma} H_{-\alpha\gamma} T(\gamma) \qquad 4.57$$

$H_{-\alpha\gamma}$ must also be linear in the applied field since these terms contribute to the vacancy wind. The matrix with elements $H_{-\alpha\gamma}$ is found from the sum of the probabilities for all defect paths from γ to $-\alpha$. For this purpose, define as before a second defect walk matrix, A_3. The elements A_{3ij} are the transition probabilities from the set of sites j into a particular site i. The matrix A_3 excludes all tracer jumps, and it excludes all jumps out of sites α for $\alpha = 1, \cdots N$. Here, however, A_3 is dependent on the field strength and hence

$$A_3 = A_{30} + \delta A_3 \qquad 4.58$$

With this definition, the matrix H with elements $H_{\delta\gamma}$ is

$$H = (I - A_3)^{-1} \qquad 4.59$$

From Eqs. 4.57–4.59, the elements $\nu_{\pi-\alpha}$ are found from

$$\nu_{\pi-\alpha} = H_{-\alpha}(I - A_3)^{-1} T \qquad 4.60$$

where $H_{-\alpha}$ is the same vector defined in Chapter 3 (see Eq. 3.79). The elements in each of the above terms can be linearized with respect to the applied field, and it follows that one has

$$\nu_{\pi-\alpha} = H_{-\alpha}(I - A_{30})^{-1} T_0 + H_{-\alpha}(I - A_{30})^{-1} \delta A_3 (I - A_{30})^{-1} T_0$$

$$+ H_{-\alpha}(I - A_{30})^{-1} \delta T = \nu_{\pi-\alpha 0} + \delta \nu_{\pi-\alpha} \qquad 4.61$$

where $\delta \nu_{\pi-\alpha}$ includes the last two terms.

Finally, the evaluation of $\delta S_+ - \delta S_-$ and S_0 in Eq. 4.51 is determined by combining Eqs. 4.47 and 4.56.

$$
\begin{aligned}
\delta S_+ - \delta S_- &= S_+ - S_- \\
&= P^{+a}[F_{(+)} - G_{(-)}] - P^{-11}[F_{(-)} - G_{(+)}] \\
&\quad - P^{-a}[F_{(-)} - G_{(+)}] + P^{+11}[F_{(+)} - G_{(-)}]
\end{aligned}
\qquad 4.62
$$

$$
\begin{aligned}
S_0 &= \tfrac{1}{2}(S_+ + S_-) \\
&= \tfrac{1}{2}\{P^{+a}[F_{(+)} - G_{(-)}] - P^{-11}[F_{(-)} - G_{(+)}] \\
&\quad + P^{-a}[F_{(-)} - G_{(+)}] - P^{+11}[F_{(+)} - G_{(-)}]\}
\end{aligned}
\qquad 4.63
$$

From Eq. 4.56, the elements $S_{+\alpha\beta}$ are found as follows:

$$
\begin{aligned}
S_{+\alpha\beta} &= Q(\beta)(I - A_2)^{-1}R(\alpha^I) - Q(\beta^I)(I - A_2)^{-1}R(\alpha^I) \\
S_{-\alpha\beta} &= Q(\beta^I)(I - A_2)^{-1}R(\alpha) - Q(\beta)(I - A_2)^{-1}R(\alpha)
\end{aligned}
\qquad 4.64
$$

As previously discussed, Q and A_2 must be linearized with respect to the field. This linearization leads to the following:

$$
\begin{aligned}
S_{+\alpha\beta} &= Q_0(\beta)(I - A_{20})^{-1}R(\alpha^I) - Q_0(\beta^I)(I - A_{20})^{-1}R(\alpha^I) \\
&\quad + [\delta Q(\beta) - \delta Q(\beta^I)](I - A_{20})^{-1}R(\alpha^I) \\
&\quad + [Q_0(\beta) - Q_0(\beta^I)](I - A_{20})^{-1}\,\delta A_2(I - A_{20})^{-1}R(\alpha^I)
\end{aligned}
\qquad 4.65
$$

The last two terms in Eq. 4.65 are the field effect terms. With a comparable equation for $S_{-\alpha\beta}$ one finds

$$
S_{0\alpha\beta} = \tfrac{1}{2}[Q_0(\beta) - Q_0(\beta^I)](I - A_{20})^{-1}[R(\alpha^I) - R(\alpha)]
\qquad 4.66
$$

$$
\begin{aligned}
(S_+ - S_-)_{\alpha\beta} &= [\delta Q(\beta) - \delta Q(\beta^I)](I - A_{20})^{-1}[R(\alpha^I) + R(\alpha)] \\
&\quad + [Q_0(\beta) - Q_0(\beta^I)](I - A_{20})^{-1}\delta A_2(I - A_{20})^{-1}[R(\alpha^I) + R(\alpha)]
\end{aligned}
$$

$$
\qquad 4.67
$$

In a very few problems, one can consider doing the calculations manually; otherwise, the equations can be programmed for computer calculation. For those few problems in which a hand calculation is feasible, Eqs. 4.66 and 4.67 are very cumbersome. In such cases, Eqs. 4.66 and 4.67 may be combined as follows:

$$
\begin{aligned}
(S_+ \pm S_-)_{\alpha\beta} &= [Q(\beta) - Q(\beta^I)](I - A_2)^{-1}R(\alpha^I) \\
&\quad \mp [Q(\beta) - Q(\beta^I)](I - A_2)^{-1}R(\alpha)
\end{aligned}
\qquad 4.68
$$

Electromigration and Thermal Diffusion of a Substitutional Solute in an fcc Metal by a Single Vacancy Mechanism

An approximate calculation of the mobility of a substitutional solute can be carried out without computer calculations. In this example N equals 1; thus the S_+ matrix is a 1×1 matrix. In the above equations, one must invert matrices associated with defect transitions, and hand calculations become cumbersome for any matrix inversion beyond 3×3. A matrix of this size is sufficient to incorporate the three classes of nearest neighbor sites to a diffusing tracer in an electric field, applied perpendicular to (100) crystallographic planes. The next approximation would include sites that are two vacancy jumps removed from the tracer; this would require a 13×13 matrix inversion, which is far too large to be considered by hand.

For this purpose define three classes of vacancy sites. Class 1 is the four sites to the right of the tracer that could lead to a tracer jump in the + field direction. Class 2 is the four-mirror image sites, and class 3 is the four sites on the same plane as the tracer. The jump frequencies $w_1 - w_4$ are the standard jump frequencies for this problem, as mentioned in the last chapter. In the field, one can write $w_i^\pm = w_i^0 (1 \pm \epsilon_i)$ associated with the direction of the atom motion in relation to the force. Hence ϵ_i is proportional to the field, and w_i^0 is not. Also, $\epsilon_i = F_i b / 2 K_B T$, where F_i is the force. Let

$$d^\pm = w_2^\pm + 2 w_1^0 + 2 w_1^\pm + 4 w_3^0 + 3 w_3^\mp$$

The calculation of the mobility requires the use of either Eqs. 4.66 and 4.67 or Eq. 4.68; equation 4.68 is used.

Introducing these jump frequencies into the defined relations in Eq. 4.65 one finds

$$[Q(\beta) - Q(\beta^I)] = 4\left(\frac{w_2^+}{d^+}, \frac{-w_2^-}{d^-}, 0\right) \qquad 4.69$$

so that

$$[\delta Q(\beta) - \delta Q(\beta^I)] = 4\left(\frac{w_2^0}{d^0} \epsilon_2 - \frac{w_2^0 \delta d^+}{(d^0)^2}\right)(1,1,0) \qquad 4.70$$

where

$$\delta d^+ = w_2^0 \epsilon_2 + 2 w_1^0 \epsilon_1 - 3 w_3^0 \epsilon_3 \quad \text{and} \quad d^0 = w_2^0 + 4 w_1^0 + 7 w_3^0$$

Also, by definition

$$R(\alpha^I) = \frac{1}{4}\begin{pmatrix} 1 \\ 0 \\ 0 \end{pmatrix}$$

and

$$R(\alpha) = \frac{1}{4}\begin{pmatrix} 0 \\ 1 \\ 0 \end{pmatrix}$$ (4.71)

Furthermore, in a manner similar with that in Chapter 3, and including the field effect, one finds

$$A_2 = \begin{pmatrix} \dfrac{2w_1^{\,0}}{d^+} & \dfrac{w_2^{\,-}}{d^-} & \dfrac{2w_1^{\,-}}{d^0} \\[2ex] \dfrac{w_2^{\,+}}{d^+} & \dfrac{2w_1^{\,0}}{d^-} & \dfrac{2w_1^{\,+}}{d^0} \\[2ex] \dfrac{2w_1^{\,+}}{d^+} & \dfrac{2w_1^{\,-}}{d^-} & \dfrac{w_2^{\,0}}{d^0} \end{pmatrix}$$ (4.72)

The vacancy wind comes partially from $\nu_{(-)}$ and one needs

$$A_3 = \begin{pmatrix} 0 & 0 & \dfrac{2w_1^{\,-}}{d^0} \\[2ex] 0 & 0 & \dfrac{2w_1^{\,+}}{d^0} \\[2ex] 0 & 0 & \dfrac{w_2^{\,0}}{d^0} \end{pmatrix}$$ (4.73)

as well as

$$H_{-\alpha} = 4(1,0,0)$$ (4.74)

and

$$T = V\begin{pmatrix} 7w_4^{\,0} & +3\epsilon_4 & w_4^{\,0} \\ 7w_4^{\,0} & -3\epsilon_4 & w_4^{\,0} \\ & 7w_4^{\,0} & \end{pmatrix}$$ (4.75)

Notice that in the absence of the field the matrices A_2, A_3, $H_{-\alpha}$, and T are the same expressions as those given in Chapter 3 for calculating the diffusivity; the superscript 0 on each w_i indicates the absence of the field. The ϵ_i is the response of the jumping atom to the applied field. For each jump frequency the response to the field may be different; hence one assumes a force F_i acts on an atom making a jump at frequency w_i. Then as previously mentioned, $\epsilon_i = F_i b/2K_B T$. For an electric field the force on the solute jump is $q_2 E$, where q_2 is the solute ionic charge and E is the field intensity. For the other jumps, all of which are solvent jumps, the force is $q_s E$, where q_s is solvent ionic charge. Thus for an electric field

$\epsilon_2 = q_2 Eb/2K_B T$ and $\epsilon_i = q_s Eb/2K_B T$ for $i \neq 2$. For a temperature gradient, one must include the chemical potential gradient of the vacancy in the motion of the solute; the temperature gradient may influence the concentration of vacancies, which in turn influences the chemical potential gradient of the vacancy. From Eq. 4.10, one would otherwise expect the force on the jumping ion to be given by $F_i = q_i(-\nabla T/T)$, where q_i is the heat of transport for a jump at frequency w_i^0. Since the chemical potential of the vacancies enters into the problem, the force for the jumping ion is given by $F_i = (q_i - \gamma \Delta H_v + E_{bi})(-\nabla T/T)$, where ΔH_v is the formation energy of the vacancy in the pure solvent, and E_{bi} is the binding energy of the vacancy to the solute at the vacancies' position prior to the jump at rate w_i. The factor γ reflects a nonequilibrium distribution of vacancies (see Eqs. 3.9 and 3.10 and the last section of this chapter). Therefore, for the temperature gradient,

$$\epsilon_1 = \frac{(q_1 - \gamma \Delta H_v + E_b)b(-\nabla T/T)}{2K_B T}$$

$$\epsilon_2 = \frac{(q_2 - \gamma \Delta H_v + E_b)b(-\nabla T/T)}{2K_B T}$$

$$\epsilon_3 = \frac{(q_3 - \gamma \Delta H_v + E_b)b(-\nabla T/T)}{2K_B T}$$

$$\epsilon_4 = \frac{(q_4 - \gamma \Delta H_v)b(-\nabla T/T)}{2K_B T} \qquad \text{4.76}$$

and for a larger than 3×3 matrix in A_2 and A_3,

$$\epsilon_0 = \frac{(q_0 - \gamma \Delta H_v)b(-\nabla T/T)}{2K_B T}$$

One may shorten the above notation by letting $q_i^* = q_i - \gamma \Delta H_v + E_{bi}$ so that the temperature gradient is

$$\epsilon_i = \frac{q_i^* b(-\nabla T/T)}{2K_B T} \qquad \text{4.77}$$

The first calculation will be the matrix S_{+0} from the matrix elements as defined in Eq. 4.66, along with the matrices in Eqs. 4.69, 4.71, and 4.72. With that substitution and some algebra,

$$S_{+0} = \frac{w_2^0}{2w_2^0 + 2w_1^0 + 7w_3^0} \qquad \text{4.78}$$

The correlation factor is related to this matrix as indicated above. Hence one finds from Eq. 4.49 that

$$f = 1 - 2S_{+0} = \frac{(2w_1{}^0 + 7w_3{}^0)}{(2w_1{}^0 + 2w_2{}^0 + 7w_3{}^0)} \qquad 4.79$$

which is in agreement with the results of the last chapter. Also, one may calculate $2\delta S_+ = (S_+ - S_-)$ from Eq. 4.67 or Eq. 4.68. One finds that

$$\delta S_+ = \{[\epsilon_2(4w_1{}^0 + 7w_3{}^0)(2w_1{}^0 + 7w_3{}^0) + 3w_3{}^0\epsilon_3(4w_1{}^0 + 7_3{}^0)$$
$$- 2w_1{}^0\epsilon_1(8w_1{}^0 + 7w_3{}^0)]/[7w_3{}^0(6w_1{}^0 + 7w_3{}^0)]\}S_{+0} \qquad 4.80$$

The frequency with which vacancies approach the tracer is calculated using Eqs. 4.61, 4.73, 4.74, and 4.75. Thus

$$\nu_{\pi-\alpha 0} = \frac{28 w_4{}^0 (6w_1{}^0 + 7w_3{}^0) V}{(4w_1{}^0 + 7w_3{}^0)} \qquad 4.81$$

and

$$\delta\nu = \nu_{\pi-\alpha 0}\left\{\frac{3\epsilon_4}{7}\frac{(4w_1{}^0 + 7w_3{}^0)}{(6w_1{}^0 + 7w_3{}^0)} - \frac{2w_1{}^0\epsilon_1}{6w_1{}^0 + 7w_3{}^0}\right\} \qquad 4.82$$

Equations 4.78, 4.80, 4.81, and 4.82 provide the basis for the calculation of the velocity due to an applied force, V_F, as indicated in Eq. 4.52 for the flux. Combining those equations using the diffusivity from the previous chapter, one finds

$$\frac{bv_F}{2D_K^*} = \epsilon_2 + \frac{\{w_3{}^0(3\epsilon_3 + 3\epsilon_4) - 4w_1{}^0\epsilon_1\}}{2w_1{}^0 + 7w_3{}^0} \qquad 4.83$$

Equation 4.83 may be used to deduce the mobility for the solute in either the case of the electric field or the temperature gradient. Thus the mobility for the electric field is

$$\frac{K_B T\bar{\mu}}{D_K^*} = q_2 + q_s \frac{(6w_3{}^0 - 4w_1{}^0)}{2w_1{}^0 + 7w_3{}^0} \qquad 4.84$$

The Einstein expression for $K_B T\bar{\mu}/D$ is q_2, thus there is an extra term included in Eq. 4.84. This extra term is due to the vacancy wind or the off-diagonal phenomenological coefficients, $L_{ij}, i \neq j$.

A more complicated expression exists for the mobility caused by a temperature gradient. Using Eqs. 4.76 and 4.83, one finds

$$\frac{K_B T\bar{\mu}}{D_K^*} = q_2^* + \frac{[(3q_3^* + 3q_4^*)w_3{}^0 - 4q_1^* w_1{}^0]}{(2w_1{}^0 + 7w_3{}^0)} \qquad 4.85$$

Thus the vacancy wind takes a slightly different form when temperature gradient is the driving force. This may be expanded to include the chemical potential of the vacancy explicitly through Eq. 4.76 as

$$\frac{K_B T \bar{\mu}}{D_K^*} = q_2 + \frac{[w_3^0(3q_3 + 3q_4 - 3E_b) - 4w_1^0 q_1]}{(2w_1^0 + 7w_3^0)}$$

$$- (\gamma \Delta H_v - E_b) \frac{(13w_3^0 - 2w_1^0)}{(2w_1^0 + 7w_3^0)} \qquad 4.86$$

RELATION OF THE OFF-DIAGONAL PHENOMENOLOGICAL COEFFICIENTS TO THE VACANCY WIND FOR A DILUTE SOLUTE

The same phenomenological coefficients appear in the flux equation when either a chemical potential gradient or the electric field is a driving force; Eq. 4.10 contains $x_j = Q_j E - \nabla \mu_j / T$, where Q_j is the effective ionic charge on ion j. As a consequence, one can use the mobility described in Eq. 4.84 to determine the vacancy wind factor, r_i, as given in Eqs. 4.31 and 4.32. For the pseudoternary alloy B, B^*, and A, described in the first part of this chapter, the flux for an electric field is given by the following:

$$J_{B^*} = L_{B^*B^*} q_B E + L_{B^*B} q_B E + L_{B^*A} q_A E \qquad 4.87$$

The forces other than the electric field are omitted; one views the specimen as being a homogeneous alloy in which an electric field has just been switched on. One may deduce the mobility of the tracer directly from Eq. 4.87 as

$$\bar{\mu}_{B^*} = \frac{J_{B^*}}{C_B^* E}$$

$$\bar{\mu}_{B^*} = \frac{q_B}{C_{B^*}} (L_{B^*B^*} + L_{B^*B}) + \frac{q_A}{C_{B^*}} L_{B^*A} \qquad 4.88$$

One can now compare Eqs. 4.84 and 4.88 for the mobility under various circumstances since the self-diffusion coefficient is known in terms of the phenomenological coefficients. That is,

$$D_B^* = \frac{K_B T L_{B^*B^*}}{C_{B^*}} \qquad 4.89$$

Combining Eqs. 4.88 and 4.89 one finds

$$\frac{K_B T \bar{\mu}_B}{D_{B^*}} = q_B \left(1 + \frac{L_{B^*B}}{L_{B^*B^*}}\right) + q_A \frac{L_{B^*A}}{L_{B^*B^*}} \qquad 4.90$$

The vacancy wind factor, r_{B^*}, is defined by Eq. 4.32, and it contains the same phenomenological coefficient ratios as Eq. 4.90.

$$r_{B*} = 1 + \frac{L_{B*B}}{L_{B*B*}} - \frac{N_B}{N_A} \frac{L_{B*A}}{L_{B*B*}} \qquad 4.32$$

From Eq. 4.90,

$$1 + \frac{L_{B*B}}{L_{B*B*}} = \frac{\partial K_B T \bar{\mu}_B / D_{B*}}{\partial q_B} \qquad 4.91$$

and

$$\frac{L_{B*A}}{L_{B*B*}} = \frac{\partial K_B T \bar{\mu}_B / D_{B*}}{\partial q_A} \qquad 4.92$$

For a dilute solute, one can rewrite Eq. 4.84 in terms of the parameters given in the above. This would imply that

$$\frac{K_B T \bar{\mu}_B}{D_B^*} = q_B + q_A \frac{(6w_3{}^0 - 4w_1{}^0)}{2w_1{}^0 + 7w_3{}^0} \qquad 4.93$$

Thus for the vacancy-solute pair model described above, Eqs. 4.91 and 4.92 would give

$$1 + \frac{L_{B*B}}{L_{B*B*}} = 1$$

so that

$$\frac{L_{B*B}}{L_{B*B*}} = 0 \qquad 4.94a$$

or

$$L_{B*B} = 0 \qquad 4.94b$$

$$\frac{L_{B*A}}{L_{B*B*}} = \frac{(6w_3{}^0 - 4w_1{}^0)}{2w_1{}^0 + 7w_3{}^0} \qquad 4.95$$

That $L_{B*B} = 0$ is easily understood. The solute is so dilute that the probability of finding solute-solute interactions is completely negligible. Interactions of this sort contribute to L_{B*B}. Also, L_{B*A} is nonzero because of the vacancy wind effect; the electric field pushes the vacancy through the solvent lattice. This alters the arrival rate of vacancies at the solute tracer; hence Eq. 4.94.

As previously mentioned, the vacancy wind factor is found in the intrinsic diffusivity, Eq. 4.32. From Eqs. 4.94 and 4.95, one finds

$$r_{B*} = 1 - \frac{N_B}{N_A} \frac{(6w_3{}^0 - 4w_1{}^0)}{(2w_1{}^0 + 7w_3{}^0)} \rightarrow 1 \qquad 4.96$$

as $N_B \rightarrow 0$. Thus for the infinitely dilute solute, the intrinsic diffusivity is found to be Eq. 4.31

$$D'_{B*} = D_{B*}\alpha$$

but $\alpha \to 1$ in that limit so that

$$D'_{B*} = D_{B*} \tag{4.97}$$

For the solvent, component A, in such a system, one can write

$$r_{A*} = 1 + \frac{L_{A*A}}{L_{A*A*}} - \frac{N_A}{N_B} \frac{L_{A*B}}{L_{B*B*}} \frac{L_{B*B*}}{L_{A*A*}} \tag{4.98}$$

by analogy with Eq. 4.32. Now

$$\frac{N_A}{N_B} \frac{L_{B*B*}}{L_{A*A*}} = \frac{D_{B*}}{D_{A*}} \tag{4.99}$$

so that one may rewrite

$$r_{A*} = 1 + \frac{L_{A*A}}{L_{A*A*}} - \frac{L_{A*B}}{L_{B*B*}} \frac{D_{B*}}{D_{A*}} \tag{4.100}$$

It seems reasonable to assume, for lack of better information at present, that

$$L_{A*B} = L_{AB} = L_{AB*} = L_{B*A}$$

so that the last term given in Eq. 4.100 is proportional to the result given in Eq. 4.95. One may determine the first two terms in Eq. 4.100 from the mobility of a solvent tracer. For a solvent tracer, $w_1^0 = w_2^0 = w_3^0 = w_4^0 = w_0^0$, $q_2 = q_s = q_A$, and this may be substituted into Eq. 4.84. Therefore, for the solvent tracer

$$\frac{K_B T \bar{\mu}_A}{D_{A*}} = q_A(1 + \tfrac{2}{9}) = \frac{q_A}{f_0} \tag{4.101}$$

when $f_0 = 9/11$, the correlation factor for the solvent self-diffusion. Now

$$1 + \frac{L_{A*A}}{L_{A*A*}} = \frac{\partial K_B T \bar{\mu}_A / D_{A*}}{\partial q_A} = \frac{1}{f_0} \tag{4.102}$$

Substituting these results into Eq. 4.100,

$$r_{A*} = \frac{1}{f_0} - \frac{D_{B*}}{D_{A*}} \frac{(6w_3^0 - 4w_1^0)}{(2w_1^0 + 7w_3^0)} \tag{4.103}$$

This final result for the vacancy wind factor, r_{A*}, gives the intrinsic diffusivity of the solvent tracer to be $\alpha = 1$,

$$D'_{A*} = D_{A*} \left\{ \frac{1}{f_0} - \frac{D_{B*}}{D_{A*}} \frac{(6w_3^0 - 4w_1^0)}{(2w_1^0 + 7w_3^0)} \right\} \tag{4.104}$$

which is in agreement with Howard and Lidiard[4]. The intrinsic diffusion coefficient of the solvent tracer differs from the self-diffusivity. The correlation factor is removed and there is a residual vacancy wind factor due to the solvent moving in a solute concentration gradient.

PHENOMENOLOGICAL DEVELOPMENT OF THE SORET EFFECT IN A BINARY SYSTEM

One cannot directly determine the heat of transport of a solute even in a dilute binary system, as given by Eqs. 4.85 or 4.86 because of vacancy wind effects. The lattice planes are moving because both the solute and solvent are mobile; their motion is driven by the temperature gradient. In any experiment with a two-component system, one must refer the fluxes to the fixed laboratory reference frame. We start with the fluxes relative to the lattice planes, as given in Eq. 4.10, with the following forces.

$$x_j = -\nabla \mu_j = \frac{-\alpha}{C_j} \nabla C_j \qquad 4.105a$$

$$x_q = -\nabla T/T \qquad 4.105b$$

and

$$x_v = \gamma \Delta \frac{H_v^0}{T} \nabla T = -\gamma \Delta H_v^0 x_q$$

$$= -\gamma \Delta H_v^0 \left(-\frac{\nabla T}{T} \right) \qquad 3.10$$

The superscript 0 denotes the average formation energy for the alloy. Then the motion of component i of the binary, with j as the other component, is from Eq. 4.10

$$J_i = \frac{-L_{ii}\alpha K_B T}{C_i} \left\{ 1 - \frac{C_i L_{ij}}{C_j L_{ii}} \right\} \nabla C_i$$

$$+ [L_{ii}(Q_i - \gamma \Delta H_v^0) + L_{ij}(Q_j - \gamma \Delta H_v^0)] x_q \qquad 4.106$$

The first term on the right-hand side is

$$-D_i^I \nabla C_i = -D_i^* \alpha r_i \nabla C_i \qquad 4.107$$

$$= -L_{ii}\alpha K_B T \left\{ \frac{1}{C_i} - \frac{L_{ij}}{C_j L_{ii}} \right\} \nabla C_i$$

It is convenient to use concentration units of mole fractions in the equation above and those that follow in this section. Also, it is convenient to define $Q_i^* = Q_i - \gamma \Delta H_v^0$, so that Eq. 4.106 may be written as

$$J_i = -D_i' \nabla N_i + [L_{ii} Q_i^* + L_{ij} Q_j^*] x_q \qquad 4.108$$

where the i stands for A or B, depending on the component of interest. The lattice planes move at a velocity

$$V_L = J_A + J_B \qquad 4.109$$

where the flux of A or B should be relative to the fixed laboratory coordinates. Thus the flux of b, relative to the fixed frame of reference is

$$J_B' = J_B - N_B V_L$$

$$= -\tilde{D}\, \nabla N_B + N_B N_A \left\{ Q_B^* \left[\frac{L_{BB}}{N_B} - \frac{L_{AB}}{N_A} \right] \right.$$

$$\left. + Q_A^* \left[\frac{L_{BA}}{N_B} - \frac{L_{AA}}{N_A} \right] \right\} x_q \qquad 4.110a$$

$$= -\tilde{D}\, \nabla N_B + \frac{N_B N_A}{\alpha K_B T} [D_B' Q_B^* - D_A' Q_A^*] x_q \qquad 4.110b$$

where $\tilde{D} = N_B D_A' + N_A D_B'$ is the mutual diffusion coefficient used earlier to describe the Kirkendall effect.

As time passes, the flux relative to the laboratory reference frame approaches zero. The contribution to the flux from the concentration gradient is balanced by the contribution from the applied forces, X_q. Therefore, $J_B' \rightarrow 0$ as $t \rightarrow \infty$, and a steady-state concentration gradient of B is established. This is called the Soret gradient, after an early worker in this field. The effect of unmixing an initially homogeneous alloy is called the Soret effect. If one sets J_B' equal to zero, from Eq. 4.110, one finds, with $X_q = \dfrac{-\nabla T}{T}$

$$\frac{\nabla N_B}{N_B} = \frac{-N_A \nabla T}{\alpha K_B T^2} \left\{ \frac{D_{B^*} r_B Q_{B^*} - D_{A^*} r_A Q_{A^*}}{N_A D_{B^*} r_B + N_B D_{A^*} r_A} \right\} \qquad 4.111a$$

$$= \frac{-N_A \nabla T}{\alpha K_B T^2} \left\{ \frac{D_B' Q_B^* - D_A' Q_A^*}{\tilde{D}} \right\} \qquad 4.111b$$

which is the Soret gradient.

Assuming the appropriate diffusivities have been measured, the measurement of the Soret gradient alone cannot determine Q_B^* since both it and Q_A^* appear in Eq. 4.111. One could also measure the lattice velocity through movement of inert markers. With Eqs. 4.109, 4.110, and algebra the lattice velocity is given by

$$V_L = -(D_A{}^I - D_B{}^I)\nabla N_B + \frac{\nabla T}{K_B T^2}\left\{N_A D_A{}^I Q_A^* + N_B D_B{}^I Q_B^*\right.$$

$$\left. + \frac{K_B T L_{BA}}{N_A N_B}\{N_B Q_B^* + N_A q_A^*\}\right\} \qquad 4.112$$

The phenomenological coefficient L_{BA} is unknown. Using the same random alloy model as with the Kirkendall effect mentioned earlier, Manning[5] finds

$$L_{BA} = \frac{2N_A N_B D_A^* D_B^*}{K_B T M_0 (N_A D_A^* + N_B D_B^*)} \qquad 4.113$$

If one makes two sets of measurements in a concentrated alloy, the lattice velocity and the Soret gradient (Eqs. 4.111 and 4.112) and if, in addition, one uses the random alloy model of Manning, then the Q_A^* and Q_B^* could be measured.

In a dilute alloy, things are somewhat easier. If the alloy is infinitely dilute, the lattice velocity of pure A becomes

$$V_L = \frac{\nabla T}{K_B T^2} D_A^* Q_A^* \qquad 4.114$$

Furthermore, in the limit of a dilute system, the Soret gradient becomes

$$\frac{\nabla N_B}{N_B} = \frac{-\nabla T}{K_B T^2}\left(Q_B^* - \frac{D_A^*}{D_B^*} r_A Q_A^*\right) \qquad 4.115$$

because $N_A \to 1$, $\alpha \to 1$, and $r_B \to 1$ from Eq. 4.96. We are left with two equations based on experimental evaluations, V_L, and $\nabla N_B/N_B$, and three unknowns Q_A^*, Q_B^* and r_A. The r_A may be found from Eq. 4.103. Therefore, an experimental measurement that gives the w_1 and w_3 are necessary to uniquely determine Q_B^*. Such a measurement is not trivial. If one definitely has a dilute solution, the grain size effect for the Soret gradient is capable of such a determination. As discussed in the next section, the grain size effect varies the coefficient γ in Eqs. 3.10, 4.76, 4.86, and so forth. The heat of transport, Q_B^*, is related to the kinetic theory value through either Eqs. 4.85 or 4.86. That is, one may equate the $KT\bar{\mu}/D_K^*$ of Eqs. 4.85 or 4.86 with the Q_B^* found in Eq. 4.115. Hence Q_B^* is an effective value that reflects many types of vacancy jumps. If one makes such a comparison, then

$$Q_B^* = q_2 + \{[w_3^0(3q_3 + 3q_4 - 3E_b) - 4w_1^0 q_1]$$
$$- (\gamma \Delta H_v - E_b)(13w_3^0 - 2w_1^0)\}/(2w_1^0 + 7w_3^0) \quad 4.116$$

Variations in γ through experimental changes in grain size could give information on w_3^0 and w_1^0.

CONCENTRATION OF VACANCIES IN A TEMPERATURE GRADIENT

The concentration of vacancies at thermal equilibrium for a single component system satisfies Eq. 3.6. It is well known that a flux of energy into such a system can displace the system from thermal equilibrium and thereby disturb the concentration of vacancies.

An extremum in the Gibb's free energy, as necessary for thermal equilibrium, requires that the system be at constant and uniform temperature and pressure. Such a situation is not present when the sample is exposed to a steady state temperature gradient. Consequently, any predictions based on an extremum in the Gibbs' function are suspect. One such prediction is that the vacancies would follow Eq. 3.6 where the temperature changes with position in the sample and the vacancy concentration follows that equation with the local temperature. This would give a position-dependent vacancy concentration. In this section we investigate the concept that the vacancy concentration might follow some position-dependent distribution other than that predicted by an extremum in the Gibbs' free energy.

According to Eq. 3.6, the fraction of vacant lattice sites would follow

$$N_v^e = \exp\left(\frac{-\Delta G_v}{K_B T}\right) \qquad 3.6$$

where it is assumed that the temperature varies along the sample. A departure from this distribution could be described by introducing a factor that changes the distribution to N_v,

$$N_v = \beta N_v^e = \beta \exp\left(\frac{-\Delta G_v}{K_B T}\right) \qquad 4.117$$

where β depends on position, and the superscript e denotes the equilibrium distribution given in Eq. 3.6.

To simplify the subsequent discussion, it is not unreasonable to presuppose that the exponential dependence is but slightly modified and assume that

$$\beta = \beta_0 \exp\,(1-\gamma)\,\frac{\Delta H_v}{K_B T} \qquad\qquad 4.118$$

where β_0 is a constant (independent of the local temperature), and γ is a function of x to be determined. Thus Eqs. 4.117 and 4.118 provide a perfectly general distribution of defects. On combining the two, one finds that

$$N_v = \beta_0 \exp\left(\frac{\Delta S_v}{K_B} - \frac{\gamma \Delta H_v}{K_B T}\right) \qquad\qquad 4.119$$

For small temperature gradients, one might expect that γ is virtually constant but not equal to 1, as would be required for an extremum in the Gibbs' function. If Eq. 4.119 is valid, then the chemical potential gradient of the vacancies follows Eq. 3.9a, which implies the validity of Eq. 3.10

$$\nabla \mu_v|_T = \gamma\,\Delta H_v\,\frac{\nabla T}{T} \qquad\qquad 3.10$$

For a single component system, one can substitute Eqs. 3.10 and 4.9 into Eq. 4.8. When this is done, and $N_v \ll 1$, one finds

$$J_1 = L_{11}(Q_1 - \gamma\,\Delta H_v)x_q \qquad\qquad 4.120a$$

$$J_q = [L_{qq} - L_{11}Q_1\gamma\,\Delta H_v]x_q \qquad\qquad 4.120b$$

where the substitutions that $x_1 = -N_v\gamma\,\Delta H_v x_q$ and $x_v = \gamma\,\Delta H_v x_q$ have been made.

The first indication that γ might differ from unity was done experimentally by Matlock and Stark[6]. They found a grain size effect in the heat of transport for pure aluminum. For single crystals the measured $Q_1 - \gamma\,\Delta H_v$ was 11.0 Kcal/mole, and for reasonably fine grain polycrystalline samples they found $Q_1 - \gamma\,\Delta H_v$ was -2 Kcal/mole. Since ΔH_v for aluminum is 17.5 kcal, they interpreted their measurements to be $\delta\gamma > 0.74$. The inequality results because L_{11} is dependent on γ as follows

$$L_{11} = \frac{C_1 D_1^*}{K_B T} = \frac{\beta C_1 D_1^{\,e}}{K_B T} \qquad\qquad 4.121$$

where β is given in Eq. 4.118 and $D_1^{\,e}$ is the diffusion coefficient assuming the equilibrium concentration of vacancies. Thus they surmised that $\gamma \cong 0$ for the single crystal sample, and $\gamma \cong 1$ for the polycrystalline sample, These values would be consistent with concentrations of vacancy sources and sinks associated with the grain boundaries.

The major departure from equilibrium thus occurs in the single crystals where γ is nearly zero. To test the hypothesis, McKee and Stark[7] measured the concentration of vacancies in aluminum through the diffusion coefficient of Ag^{110} in Al. They found that the diffusivity of silver at the cold end of a temperature gradient in aluminum satisfies

$$D_1^* = 4b^2 \frac{w_4}{w_3} w_2 f \exp\left\{ \frac{\Delta S_v}{k} - (1-\gamma)\frac{\Delta H_v}{KT_H} - \frac{\gamma\Delta H_v}{KT} \right\} \qquad 4.122$$

$$= \beta D_1^e$$

where, experimentally, $\gamma = 0.35$ in single crystalline aluminum for the $\nabla T \neq 0$. Their results constitute proof that γ can depart from 1, since in their experiments, β varied between 2 and 6, which was far beyond experimental error.

Theoretically, one can predict that γ should depart from unity using theorems of irreversible thermodynamics. The first such theorem is one deduced by Prigogine, which minimizes the rate of entropy production in a steady-state irreversible process. The rate of entropy production for a local volume element of the sample is given by Eq. 2.85. When one eliminates the vacancies as a component, as was done in the early part of this chapter, the entropy production equation, 2.85, becomes for a single component system

$$T\sigma = J_q x_q + J_1 (x_1 - x_v) \qquad 4.123$$

On substituting Eqs. 4.20 and 4.21 into Eq. 4.123 one finds that

$$T\sigma = \left\{ L_{qq} - \frac{\beta_0 C_1 D_1^e}{K_B T} \gamma \Delta H_v (2Q - \gamma \Delta H_v) \exp(1-\gamma)\frac{\Delta H_v}{K_B T} \right\} x_q^2 \qquad 4.124$$

The minimum rate of entropy production theorem asserts that

$$\delta \int_v T\sigma \, dv = 0 \qquad 4.125$$

where the δ denotes the variation, and v is the volume.

Equation 4.125 is satisfied when

$$\frac{\partial T\sigma}{\partial \gamma} = 0$$

and

$$\frac{\partial^2 T\sigma}{\partial \gamma^2} > 0$$

where $T\sigma$ is given by Eq. 4.124. With some algebra, Stark[8] finds that these two conditions are satisfied when

$$\gamma = \frac{Q}{\Delta H_v} + \frac{K_B T}{\Delta H_v} - \left\{ \left(\frac{Q}{\Delta H_v} \right)^2 + \left(\frac{K_B T}{\Delta H_v} \right)^2 \right\}^{1/2} \qquad 4.126$$

for $Q > 0$; the minus sign becomes a plus for $Q < 0$. For aluminum the $Q > 0$ was implied by the measurements of Matlock and Stark, and the ΔH_v measured by Simmons and Balluffi. Then, one finds that $\gamma = 0.08$ when these values are substituted into Eq. 4.126. This value may be compared to $\gamma = 0.35$,[9] determined experimentally. Evidently, the minimum rate of entropy production does not give an extremely good value for γ. It is noteworthy, however, that the minimum rate of entropy production does predict a value of γ much closer to that observed than does the assumption that $\gamma = 1$, as predicted by a minimum local Gibbs free energy.

Matlock and Stark observed the length of a single crystal sample increasing in the temperature gradient. Such an observation is consistent with a large flux of mass at the hot end flowing towards the cold end. A small flux of mass at the cold end also flows toward the cold end with $Q_1 > 0$ in this case. The sample then lengthens, since the flux at the hot end is larger in magnitude than that at the cold end.

On the other hand, if the lateral surface of the sample acts as a sink for vacancies, the vacancies could have been produced at the hot end and flowed toward the cold end in addition to the lateral surface, $Q_v = -Q_1 > 0$. Then, since the lateral surface is only acting as a sink for the vacancies, the specimen contracts in that direction and lengthens along the gradient. This is again consistent with the Matlock and Stark observation of specimen lengthening. Matlock and Stark made measurements of changes in one of the lateral dimensions. Their results were negative; there are two lateral dimensions, one may change leaving the other almost constant.

The two possible explanations can be summarized as follows: (a) The vacancies flow from the surface via dislocations in a manner that does not conserve lattice sites into the interior where they appear to be formed or they are created at dislocations; then they flow toward the hot end of the sample and the lateral surface giving rise to the appearance of an effective internal source. (b) The vacancies are created at the hot end and they flow toward the cold end and toward the lateral surface with literally no internal sources and sinks operating. In either case the concentration of vacancies is not the value that is given by thermal equilibrium, as is evidenced by the value of $\gamma = 0.35$ determined by diffusion measurement from the cold end of the sample.[10]

Explanation (a) requires an internal vacancy source term to be proportional to $(\nabla T)^2$; this suggests the entropy production since the terms in it are proportional to $(\nabla T)^2$. On the other hand, if the surfaces of the sample are at equilibrium, the vacancies in the interior are not at equilibrium without internal sinks. This can be deduced quite easily by using a one-dimensional model. Therein, one assumes no vacancy sources or sinks interior to the crystal. The hot and cold ends are the source and sink for vacancies, respectively. Then, one has the following for the steady-state vacancy flux with no sources:

$$\frac{d}{dx} J_v = 0; \qquad J_v = a = \text{constant} \qquad\qquad 4.127$$

where

$$J_v = -D_v \left[\frac{dN_v}{dx} + \frac{N_v Q_v \nabla T}{KT^2} \right] \qquad\qquad 4.128$$

Assuming that the hot and cold ends of the sample are at equilibrium with respect to vacancies concentration, one finds the following:

$$D_v = D_{0v} e^{-\Delta H_m/KT}$$

$$\beta = \frac{a(Q - \Delta H_m)}{D_{0v} K |\nabla T|}$$

and

$$N_v = \exp \frac{Q}{K}\left(\frac{1}{T} - \frac{1}{T_H}\right) \left\{ \exp - \frac{\Delta G_v}{KT_H} - \beta \exp \frac{Q}{KT_H} \times \right.$$
$$\left. \left[\left(\frac{e^{-U_2}}{U_2} - E_1(U_2)\right) - \left(\frac{e^{-U_1}}{U_1} - E_1(U_1)\right) \right] \right\} \qquad 4.129$$

where

$$U_1 = \frac{KT_H}{Q - \Delta H_m}$$

$$U_2 = \frac{KT}{Q - \Delta H_m}$$

and

$$E_1(x) = \int_x^\infty \frac{e^{-t}}{t} \, dt$$

In Eq. 4.129, the only condition that yields vacancy equilibrium at any position in the sample is for $Q = -\Delta H_v$, in which case $\beta = 0$ because $a = 0$. Otherwise, the surface cannot maintain vacancy equilibrium within the interior, and with $Q > -\Delta H_v$ one has an overequilibrium concentration of vacancies in the crystal, which is consistent with the experiments.

The dilemma in the two explanations is: to accept explanation (a), which requires internal sources operating, one must define some model where the vacancy source strength is proportional to $(\nabla T)^2$. This has yet to be accomplished. On the other hand, to accept explanation (b) requires the inoperation of nearly all internal vacancy sources and sinks: the surfaces are the only operable ones. Nevertheless, the specimen lengthens uniformly, which requires the production of internal atom planes; such a process evidently requires dislocation climb in contrast to explanation (b).

REFERENCES

1. J. R. Manning, *Diffusion Kinetics for Atoms in Crystals*, D. Van Nostrand, Princeton, N. J., 1968.
2. J. P. Stark and J. R. Manning, *Phys. Rev.* **89,** 425 (1974).
3. R. E. Howard, *Phys. Rev.* **144,** 650 (1965).
4. R. E. Howard and A. B. Lidiard, *J. Phys. Soc. Japan* **18,** 197 (1963); supplement II.
5. J. R. Manning, ibid.
6. J. H. Matlock and J. P. Stark, *Acta Met.* **19,** 923 (1971)
7. R. A. McKee and J. P. Stark, *Phys. Rev.* **B11,** 1374 (1975)
8. J. P. Stark *Scripta Net.* **5,** 727 (1971), there is a correction in Eq. 4.126 over that given by Stark in the original work.
9. R. A. McKee and J. P. Stark, ibid.
10. R. A. McKee and J. P. Stark, *Phys. Rev.* **B11,** 1374 (1975).

CHAPTER 5

DIVACANCIES AND CONCENTRATION EFFECTS IN DILUTE ALLOYS

THE CORRELATION FACTOR

The motion of a solute tracer in a dilute alloy, influenced by a vacancy mechanism (as described in Chapters 3 and 4), is complicated by the probability that a second vacancy or a second solute atom can influence the correlated tracer motion. For example, when two vacancies are simultaneously moving a tracer, the jump rates are significantly increased. The increase in jump rate can be sufficiently large that the number of tracer jumps with vacancy pairs is comparable with that of isolated vacancies, even though the probability that one observes vacancy pairs is small. The ability to deal with this type of complication was introduced in the tracer mobility in the last chapter, even though it was not necessary for the problem at hand. The formalism was presented as a set of matrix equations that could be reduced to a scalar for the single vacancy-solute problem to be developed therein. The same complication, that of a set of distinguishable jump configurations, is present in the diffusivity and correlation factor. The necessary matrix equations are now developed. To begin, consider Eq. 3.30 for the correlation factor. This equation was reduced to Eq. 3.34 by assuming that $\langle x_i x_{i+j} \rangle = \langle x_1 x_{1+j} \rangle$ for all i. When one has distinguishable jumps, this simplification cannot be made. One must take an alternate route; Eq. 3.30 gives the correlation factor as

$$f = \lim_{n \to \infty} \left\{ 1 + \frac{2}{n} \sum_{j=1}^{n-1} \sum_{i=1}^{n-j} \frac{\langle x_i x_{i+j} \rangle}{b^2} \right\} \qquad 3.30$$

Herein b is the x projection of the jump trajectory. When there are a set of distinguishable jumps, α, $\alpha = 1, \cdots N$, then one must consider the probability that the ith jump is of type α so that $\langle x_\alpha x_{\alpha j} \rangle = \langle x_i x_{i+j} \rangle$ for a particular value of i. This could also happen for any value of i. To circumvent notational difficulty, one introduces the fraction of jumps of type α to be C_α. Then C_α will be the probability that any jumps $\langle x_i x_{i+j} \rangle$ will actually be an α jump; and one is concerned with the jth jump following an α jump, $\langle x_\alpha x_{\alpha,j} \rangle$. Thus instead of considering the jth jump following an ith jump we consider the jth jump following an α type jump, which could be a γ type jump, $\gamma = 1, \cdots N$. The probability of this is C_γ. Hence one can calculate the correlation factor by the equation developed by Howard[1]

$$f = 1 + 2 \sum_{\alpha=1}^{N} C_\alpha \sum_{j=1}^{\infty} \frac{\langle x_\alpha x_{\alpha,j} \rangle}{b^2}$$

$$= \sum_{\alpha=1}^{N} C_\alpha \left\{ 1 + 2 \sum_{j=1}^{\infty} \frac{\langle x_\alpha x_{\alpha,j} \rangle}{b^2} \right\} \qquad 5.1$$

since $\sum_{\alpha=1}^{N} C_\alpha = 1$. The term in braces would be identical to the correlation factor if α were restricted to one type of jump, as in the case for the isolated tracer-vacancy pair. Thus when N differs from unity, one is tempted to introduce the notion of a partial correlation factor, f_α, which is associated with an α type jump. Hence one defines

$$f_\alpha = 1 + 2 \sum_{j=1}^{\infty} \frac{\langle x_\alpha x_{\alpha,j} \rangle}{b^2} \qquad 5.2a$$

so that

$$f = \sum_{\alpha=1}^{N} C_\alpha f_\alpha \qquad 5.2b$$

Regarding the partial correlation factors and the terms $\langle x_\alpha x_{\alpha,j} \rangle$, one must be concerned with the probability that the jth jump following an α jump is of type β, and is either parallel (11) or antiparallel (a), with jump α. Therefore, as in the last chapter but excluding any applied field, $P_{\alpha\beta}^{j11}$ is the probability that on the jth jump following a jump of type α a β type jump parallel to jump α will result. Furthermore, $P_{\alpha\beta}^{ja}$ is the probability that the jth jump following an α jump is of type β, and antiparallel to jump α. With these probabilities,

$$\langle x_\alpha x_{\alpha,j} \rangle = \sum_{\beta=1}^{N} |x_\alpha|\{+|x_{\alpha,j}|P_{\alpha\beta}^{j11} + (-|x_{\alpha,j}|P_{\alpha\beta}^{ja})\}$$

$$= b^2 \sum_{\beta=1}^{N} (P_{\alpha\beta}^{j11} - P_{\alpha\beta}^{ja}) \qquad 5.3$$

One may develop a recursion formula for these probabilities, in the same manner as was done in Chapter 4, by considering $\langle x_\alpha x_{\alpha,j+1} \rangle / b^2$. By definition one has

$$\langle x_\alpha x_{\alpha,j+1} \rangle = b^2 \sum_{\beta=1}^{N} (P_{\alpha\beta}^{j+111} - P_{\alpha\beta}^{j+1a}) \qquad 5.4$$

or

$$\frac{\langle x_\alpha x_{\alpha j+1} \rangle}{b^2} = \sum_{\beta=1}^{N} \sum_{\gamma=1}^{N} \{(P_{\alpha\gamma}^{j11} P_{\gamma\beta}^{111} + P_{\alpha\gamma}^{ja} P_{\gamma\beta}^{1a})$$
$$- (P_{\alpha\gamma}^{j11} P_{\gamma\beta}^{1a} + P_{\alpha\gamma}^{ja} P_{\gamma\beta}^{111})\}$$
$$= \sum_{\beta=1}^{N} \sum_{\gamma=1}^{N} \{P_{\alpha\gamma}^{j11}(P_{\gamma\beta}^{111} - P_{\gamma\beta}^{1a})$$
$$- P_{\alpha\beta}^{ja}(P_{\gamma\beta}^{111} - P_{\gamma\beta}^{1a})\} \qquad 5.5$$

It is apparent from Eq. 5.5 that one may deduce $P_{\alpha\beta}^{j+111} - P_{\alpha\beta}^{j+1a}$ from the following:

$$P_{\alpha\beta}^{j+111} - P_{\alpha\beta}^{j+1a} = \sum_{\gamma=1}^{N} (P_{\alpha\gamma}^{j11} - P_{\alpha\gamma}^{ja})(P_{\gamma\beta}^{111} - P_{\gamma\beta}^{1a}) \qquad 5.6$$

Consequently, if one defines an $N \times N$ matrix P_2^j with elements given as

$$P_2^j = P_{\alpha\beta}^{j11} - P_{\alpha\beta}^{ja} \qquad 5.7$$

Eq. 5.6 defines a matrix multiplication whereby

$$P_2^{j+1} = P_2^j P_2^1 \qquad 5.8$$

Evidently, by induction

$$P_2^j = (P_2^1)^j = (P_2)^j \qquad 5.9$$

Comparing Eqs. 5.3, 5.7, and 5.9, one finds that the matrix elements $\langle x_\alpha x_{\alpha,j} \rangle / b^2$ are found in the vector as follows:

$$\frac{\langle x_\alpha x_{\alpha,j} \rangle}{b^2} = P_2^j 1 \qquad 5.10$$

where 1 is the N-dimensional unit column vector; the post multiplication by 1 sums over β. Substituting Eq. 5.10 into Eq. 5.2a gives the partial correlation factors, f_α, as elements of a column vector f defined by:

$$f_* = 1 + 2P_2(I - P_2)^{-1}1 \qquad 5.11$$

where I is the N-dimensional unit matrix.

Then if one were to write a row vector $\underline{C}$ whose elements are C_α, the equation for the correlation factor becomes

$$f = \underline{C}f_* \qquad 5.12$$

One may also write a correlation factor matrix $\underline{f}$ which is $N \times N$. $\underline{f}$ is defined by the relationships

$$f_* = \underline{f}1 \qquad 5.13a$$

so that

$$\underline{f} = I + 2P_2(I - P_2)^{-1}$$
$$= (I + P_2)(I - P_2)^{-1} \qquad 5.13b$$

Note that P_2 is a generalization of that defined in Chapter 3 to an $N \times N$ matrix. This matrix $\underline{f}$ was identified in the last chapter in Eq. 4.49 in terms of the matrix $\bar{S}_{0+}$.

CALCULATION OF THE FRACTION OF JUMPS, C_α, AND THE AVERAGE JUMP FREQUENCY

Since there are N jump configurations, $\alpha = 1, \cdots N$, for the solute tracer, it is apparent that there will exist a given distribution of these complexes such that one can define the steady-state probability that the tracer is found in configuration α. Suppose that by some means one can calculate the number of ways, M_α, that the tracer in a given configuration, α, can be found. Also, suppose that when the tracer is in configuration α its jump frequency is W_α. Given these two numbers, one can calculate the average jump frequency of the tracer, $\langle W \rangle$. It will be the sum over the probability, U_α, that the tracer jumps at a rate W_α. Evidently, since there are at least $N+1$ configurations where $\alpha = 1, \cdots N$, $U_\alpha = M_\alpha \Big/ \sum_{\beta=1}^{N+1} M_\beta$, so that the average frequency of tracer jumps may be determined from

$$\langle W \rangle = \sum_{\alpha=1}^{N} U_\alpha W_\alpha = \sum_{\alpha=1}^{N} M_\alpha W_\alpha \Big/ \sum_{\beta=1}^{N+1} M_\beta \qquad 5.14$$

The $N+1$st configuration is not a jump configuration; the tracer has no neighboring defect. The U_α is also the fraction of solute atoms that are contained in configuration α. By definition, the C_α is the fraction of jumps the tracer makes at a rate W_α, which from Eq. 5.14 must be.

$$C_\alpha = \frac{U_\alpha W_\alpha}{\langle W \rangle} \qquad 5.15$$

Finally, the diffusion coefficient may be written using the concept above. For an fcc and bcc crystal, there are eight positions that lead to a tracer jump along the x axis of magnitude $\sqrt{2}b$; b is the x projection of the jump trajectory. This leads to

$$D_K^* = \frac{\langle x^2 \rangle}{2t} = \frac{\langle \rho^2 \rangle}{6t} = \frac{b^2}{3} \langle W \rangle f \qquad 5.16$$

where the three-dimensional jump frequency, $\langle W \rangle$, is found in Eq. 5.14 and f is the combination of Eqs. 5.11, 5.12, and 5.15.

To calculate the diffusivity, one needs a technique to determine the functions U_α; this can be done using either statistical mechanics or vacancy annealing kinetics. An example of the vacancy annealing kinetics is the calculation of the C complex in Chapter 3, Eq. 3.16. One could also use the random walk approach to calculate the diffusivity through the matrices $P_{\alpha\beta}^{11}$ and $P_{\alpha\beta}^{a}$, as in Chapter 3. Examples of these methods are given below.

SOLUTE ENHANCEMENT OF SOLVENT SELF-DIFFUSION

The vacancy jump frequencies in a dilute alloy have been viewed as being altered due to the presence of a solute atom. The jump rates, w_i, $i = 1, \cdots 4$, are different from w_0. In the pure material, the solvent tracer jumps only at the rates w_0. When one introduces the dilute solute, some of the solvent tracer jumps will be at the frequencies w_1, w_3, and w_4. As a consequence, the average jump frequency of the solvent tracer is changed. When the jump frequency is changed, the correlation factor is also expected to be changed. We investigate the details of this in what follows: For the specific example, we parallel the developments of Lidiard[2] and of Howard and Manning[3].

In a face-centered cubic lattice the jump frequencies of a solvent tracer atom can be changed from w_0 when the tracer is found in the first- or second-neighbor shell to the solute atom. When the solvent tracer is on the first-neighbor shell, it can jump at a rate w_1 and w_4. When the solvent tracer is on the second-neighbor shell it can jump at a rate w_3 and w_0. We calculate the number of jumps that are altered by the frequencies that differ from w_0; we view the process as if we were sitting on the solvent tracer. The probability that the solute and vacancy are on a particular pair of sites is $I_0 V$ for second-neighbor pairs and $w_4 I_0 V/w_3$ for first-neighbor pairs. There are 42 second-neighbor pairs for the fcc lattice.

Consider the tracer on the first-neighbor shell to the solute. There are four sites the tracer could jump to at a frequency w_1 with a probability of $12(w_4/w_3)I_0 V$. There are seven sites the tracer could jump to at a frequency w_4 with a probability of $12 I_0 V$.

If the tracer is on the second-neighbor shell, there are seven tracer sites that could exchange with a vacancy at a particular site on the first-neighbor shell, at a frequency w_3, with a probability $12 I_0 V w_4/w_3$. Thus the frequency is changed from w_0 for the following jumps

$$4w_1[12(w_4/w_3)I_0 V] + 7w_4[12 I_0 V] + 7w_3[12(w_4/w_3)I_0 V]$$
$$= [4w_1 + 14w_3]12 I_0 V w_4/w_3 \quad 5.17$$

These frequencies replace jumps at rate w_0. The total number of jumps at rate w_0 that are altered are obtained from Eq. 5.17 by letting $w_1 = w_4 = w_3$. Hence the change in jump frequency $\delta\langle W\rangle$ due to the presence of the solute atom is

$$\delta\langle W\rangle = 12 I_0 V(w_4/w_3)[4w_1 + 14w_3 - 18] \quad 5.18$$

When $I_0 = 0$, the solvent tracer jump frequency is

$$\langle W(I_0 = 0)\rangle = 12 V w_0 \quad 5.19$$

The average jump frequency with the solvent present is then the sum of frequencies without the solute and the change in the frequency due to the presence of the solute. Thus $\langle W\rangle$ is found from the sum of Eqs. 5.18 and 5.19.

$$\langle W\rangle = 12 V w_0 \left\{ 1 - 18 I_0 + w_4 I_0 \frac{4w_1 + 14w_3}{w_3 w_0} \right\} \quad 5.20$$

It is convenient for calculating the correlation factor to rewrite this equation letting the vacancy fraction remain V, and letting $C = 12 V I_0 w_4/w_3$. One can then write the average frequency as

$$\langle W\rangle = 12 \left\{ V(1 - 18 I_0)w_0 + \frac{C}{3}[w_1 + \tfrac{7}{2}w_3] \right\} = 12\gamma \quad 5.21$$

The jump frequency times the correlation factor is needed in the diffusivity. When the tracer is at least a second neighbor to the solute, one may assume that the partial correlation factor for jumps at rate w_0 is unaffected by the presence of the solute. Thus one can write

$$\langle W \rangle f = \langle W(I_0 = 0) \rangle f_0 \left[(1 - 18 I_0) + \frac{C}{3}(w_1 X_1 + \tfrac{7}{2} w_3 X_2) \right] \qquad 5.22$$

where $f_0 = 0.781$ for the face-centered cubic structure. Hence one may write

$$\langle W \rangle f = \langle W(I_0 = 0) \rangle f_0 \{1 - B I_0\} \qquad 5.23$$

where

$$B = -18 + \frac{4 w_4}{f_0 w_0} \left(\frac{X_1 w_1}{w_3} + \frac{7 X_2}{2} \right) \qquad 5.24$$

The factors X_1 and X_2 are related to the partial correlation factors. The notational changes introduced by Eqs. 5.21–5.24 are convenient for computer calculation and simplifications of the final form. Explicitly, one must identify the specific jump configurations α, $\alpha = 1$, N, calculate the partial correlation factors, calculate the C_α, and then relate the partial correlation factors and C_α to product $\langle W \rangle f$. When this is done, one may determine the X_1 and X_2.

Some of the work outlined in the previous paragraph has been accomplished because of the specific final form for Eqs. 5.23 and 5.24. Actually only the partial correlation factors and their relationship to X_1 and X_2 are left to computer calculation. However, to understand how this comes about, one must detail the calculation of the partial correlation factors, the C_α, and so forth.

First, one must identify the various tracer jump configurations. To do this let the initial tracer position prior to the jump be the origin, $b(0, 0, 0)$, and the initial vacancy position be $b(1, 1, 0)$. We then identify the possible positions of the impurity for a possible tracer-vacancy exchange with the X axis perpendicular to (100) planes. With these specifications, the jump frequencies, C_α, and configurations are specified by Table 5.1. It is important to notice that the C_α for $\alpha = 1$ and 2 are proportional to w_1, whereas the C_α, $\alpha > 2$ are proportional to w_3. Because of this one may collect the partial correlation factors into two groups and average them. Thus one lets X_1 and X_2 be the averages of the partial correlation factors for these two groups. These averages are

$$X_1 = \tfrac{1}{2}(f_1 + f_2)$$
$$X_2 = \tfrac{1}{14}(f_3 + 2 f_4 + f_5 + 2 f_6 + f_7 + f_8 + 2 f_9 + f_{10} + 2 f_{11} + f_{12}) \qquad 5.25$$

The factors of 2 found in Eq. 5.25 came from the C_α as given in Table 5.1. The calculation of B in Eq. 5.24 and hence the diffusivity is reduced

Table 5.1 Jump configurations[a]

Type	Impurity position	Jump frequency	$C_\alpha\gamma$[b]
1	$b(1,0,1)$	W_1	$CW_1/6$
2	$b(0,1,1)$	W_1	$CW_1/6$
3	$b(2,0,0)$	W_3	$CW_3/12$
4	$b(2,1,1)$	W_3	$CW_3/6$
5	$b(2,2,0)$	W_3	$CW_3/12$
6	$b(1,2,1)$	W_3	$CW_3/6$
7	$b(0,2,0)$	W_3	$CW_3/12$
8	$b(1,-1,0)$	W_4	$CW_3/12$
9	$b(0,-1,1)$	W_4	$CW_3/6$
10	$b(-1,-1,0)$	W_4	$CW_3/12$
11	$b(-1,0,1)$	W_4	$CW_3/6$
12	$b(-1,1,0)$	W_4	$CW_3/12$

[a] γ is given by Eq. 5.21; $\gamma = \langle W \rangle /12$
[b] R. E. Howard and J. P. Manning, ibid.

to the calculation of the partial correlation factors, f_α, and their collection into averages, as given by Eq. 5.25. The diffusivity will then be found through the combination of Eqs. 5.16, 5.23, 5.24, and 5.25. Hence one can write the diffusion coefficient for the solvent tracer as

$$D = D(I_0 = 0)(1 - BI_0) \qquad 5.26$$

where B is given in Eq. 5.24. The $D(I_0 = 0)$ is the usual expression for the self-diffusion coefficient of a tracer in a pure material. The B requires the tabulation of X_1 and X_2, which must be done by random walk since they involve the partial correlation factors.

The partial correlation factors are found as the elements of a 12-component column vector as given in Eq. 5.11, with P_2^1 given by Eqs. 5.7 and 5.9. Thus, on deleting the one as a superscript, P_2 is a 12×12 matrix. Now one may calculate the matrix elements

$$P_2^{\alpha\beta} = P_{\alpha\beta}^{11} - P_{\alpha\beta}^a \qquad 5.27$$

directly, or one may take a larger matrix and calculate $P_{\alpha\beta}^{11}$ and $P_{\alpha\beta}^a$. The larger matrix method uses the matrix element calculation where matrix A is defined as in Chapter 3, and the vectors $Q(\beta)$ and $R(\alpha)$ are defined as in Chapter 4. Thus the elements of the matrix are

$$P_{\alpha\beta}^{11} = Q(\beta)(I - A)^{-1} R(\alpha)$$

and

$$P_{\alpha\beta}^a = Q(\beta)(I-A)^{-1}R(\alpha^I) \qquad 5.28$$

The matrix A includes a set of configurations $M > N$, the mirror image of these configurations, plus all configurations where the solute, tracer, and vacancy are contained on the same X plane. Thus the matrix A is of dimensionality $2M + J$, when there are J configurations where all these species are on the same X plane.

In practice, this is overly cumbersome and one may calculate $P_2^{\alpha\beta}$ directly by reducing the dimensionality of A to $M \times M$, so that A only contains those configurations where the vacancy is on one side of the tracer. Then

$$P_{\alpha\beta}^a - P_{\alpha\beta}^{11} = Q(\beta)(I-A)^{-1}R(\alpha^I) \qquad 5.29$$

Image configurations (i^I) are treated as the same configuration (i) in Eq. 5.29. Since the vacancy cannot travel to the other side of the tracer, the calculation in Eq. 5.29 automatically excludes those vacancy paths where the vacancy finds itself on the same plane as the tracer, from which it would equally well contribute to P^{11} or P^a. Howard and Manning[4] use this latter method with the 12-jump configurations mentioned in Table 5.1, the 9-jump configurations where the solute separates the vacancy from the tracer, and 98 additional jump configurations. They therefore use a matrix, A, of dimension 119×119. The row vector $Q(\beta)$ is 1×119, and the column vector $R(\alpha^I)$ is 119×1. They find that when the w_i are all equal, $X_1 = X_2 = 0.79$, which is comparable to the exact answer of 0.781. Otherwise, $0 < X_1 < 1$, and $0.3 < X_2 < 1$, and X_1 and X_2 depend on the jump frequency ratios w_3/w_1, w_2/w_1, and w_4/w_1. The w_2 enters the problem as a result of solute vacancy jumps. The reader is referred to their paper for tabulation of X_1 and X_2. For reference, similar tabulations are given in the next problem, as well as some explicit matrix elements.

Once the matrix elements within P_2 are available, one must calculate the partial correlation factor using Eq. 5.11. One therefore must invert two matrices, an $M \times M$ matrix of configurations plus the $N \times N$ matrix when the other is accomplished. Alternatively, one could calculate the S_{0+} matrix by Eq. 4.66, from which one may calculate the $\underline{f}$ matrix through Eq. 4.49. Thus

$$\underline{f} = I - 2S_{0+} \qquad 4.49$$

The partial correlation factors are then found by Eq. 5.13a as elements of the column vector f_*. This latter procedure requires the single matrix inversion of $(I - A_2)^{-1}$, where A_2 is $2M + J$ dimensionality. This latter procedure is more convenient when the mobility is being calculated, as shown later in this chapter.

INFLUENCE OF SOLUTE PAIRS ON SOLUTE SELF-DIFFUSION IN AN fcc LATTICE[5]

When a second solute atom is in the first- or second-neighbor shell to a solute tracer, the jump frequency of the tracer will be altered. This gives a concentration dependence to the solute self-diffusion coefficient; this section considers such a dependence for an fcc lattice. The concentration of various tracer-solute-vacancy complexes follows a minimum in free energy. The minimum in free energy is a manifestation of the vacancy annealing kinetics. Equilibrium in the system would reflect the proper forward and reverse reactions for sources and sinks of vacancies. If one assumes that the solute concentration is low, then these sources and sinks reflect the production and annihilation of unassociated vacancies. Such an approach avoids a temperature dependence of the solute concentration at these vacancy sources and sinks. Furthermore, the divacancy is neglected; however, pairs of two solute atoms are not.

The following definitions are pertinent:

$I_0 =$ atom fraction solute.

$I =$ atom fraction unassociated solute atoms.

$I_2 =$ atom fraction of solute pairs (no vacancy present).

$J_0 =$ atom fraction of solute pairs with an associated vacancy at 60°.

$J_1 =$ atom fraction of solute pairs with an associated vacancy at 90°.

$J_3 =$ atom fraction of solute pairs with an associated vacancy at 180°.

$J_5 =$ atom fraction of solute pairs with an associated vacancy at 120°.

$J_2 =$ atom fraction of solute-vacancy pairs with an associated solute at 90°.

$J_4 =$ atom fraction of solute-vacancy pairs with an associated solute at 180°.

$J_6 =$ atom fraction of solute-vacancy pairs with an associated solute at 120°.

$J =$ $J_2 + J_4 + J_6$.

$K_i =$ jump rate for one of the reactions.

$L_i =$ reaction rate.

$L =$ reduced reaction rate.

$M =$ $J_0 + J_1 + J_3 + J_5$

$N_i =$ configurational coefficient.

$V =$ unassociated vacancy fraction.

$w_1, w_2, w_3, w_4 =$ usual jump frequencies associated with correlation factor.

Consistent with this, one must consider the following reactions.

$$J_2 \xrightleftharpoons[L_2]{L_1} C + I$$

$$J_4 \xrightleftharpoons[L_4]{L_3} C + I$$

$$J_6 \xrightleftharpoons[L_6]{L_5} C + I$$

$$J_0 \xrightleftharpoons[L_8]{L_7} I_2 + V$$

$$J_1 \xrightleftharpoons[L_{10}]{L_9} I_2 + V$$

$$J_3 \xrightleftharpoons[L_{12}]{L_{11}} I_2 + V$$

$$J_5 \xrightleftharpoons[L_{14}]{L_{13}} I_2 + V$$

$$J_2 \xrightleftharpoons[L_{16}]{L_{15}} 2I + V$$

$$J_4 \xrightleftharpoons[L_{18}]{L_{17}} 2I + V$$

$$J_6 \xrightleftharpoons[L_{20}]{L_{19}} 2I + V$$

$$J_0 \xrightleftharpoons[L_{22}]{L_{21}} J_1$$

$$J_1 \xrightleftharpoons[L_{24}]{L_{23}} J_2$$

$$J_1 \xrightleftharpoons[L_{26}]{L_{25}} J_5$$

$$J_3 \xrightleftharpoons[L_{28}]{L_{27}} J_4$$

$$J_3 \xrightleftharpoons[L_{30}]{L_{29}} J_5$$

$$J_5 \xrightleftharpoons[L_{32}]{L_{31}} J_6$$

$$J_0 \xrightleftharpoons[L_{34}]{L_{33}} J_5$$

$$C \xrightleftharpoons[L_{36}]{L_{35}} I + V \qquad\qquad 5.30$$

Also, there are some internal conversion reactions that are important in the calculation of the correlation factor. These are

$$J_0 \xrightarrow[K_{37} \text{ (solvent jump)}]{K_0 \text{ (solute jump)}} J_0$$

$$J_1 \xrightarrow{K_{38}} J_1$$

$$J_5 \xrightarrow{K_{39}} J_5$$

$$J_2 \xrightarrow{K_{40}} J_2$$

$$J_6 \xrightarrow{K_{41}} J_6 \qquad\qquad 5.31$$

In some systems, divacancies may be important. In such a system, it is likely that they will be as important at infinite dilution as at a small finite solute concentration, as in this model. The divacancy contribution should be studied first at infinite dilution (see the next section for that analysis). The interaction between two vacancies and two solute atoms introduces an extreme complication, which is unsolved. The contribution of divacancies in pure solvent self-diffusion is questionable, but appears to be important in silver self-diffusion.

Writing the rates as $L_i = N_i K_i$, K_i is the jump frequency and N_i is the number of ways for the jump to occur. One finds the following:

$N_1 = 4$	$N_{19} = 3$
$N_2 = 4 \times 3 \times 4 = 48$	$N_{20} = 3 \times 2 \times 4 \times 4 = 96$
$N_3 = 8$	$N_{21} = 2$
$N_4 = 8 \times 6 = 48$	$N_{22} = 2$
$N_5 = 6$	$N_{23} = 1$
$N_6 = 6 \times 2 \times 4 \times 4 = 192$	$N_{24} = 2$
$N_7 = 5$	$N_{25} = 2$
$N_8 = 5 \times 4 \times 6 = 120$	$N_{26} = 1$
$N_9 = 6$	$N_{27} = 1$
$N_{10} = 6 \times 4 \times 6 = 144$	$N_{28} = 2$
$N_{11} = 7$	$N_{29} = 4$

$$N_{12} = 7 \times 2 \times 6 = 84 \qquad N_{30} = 1$$
$$N_{13} = 7 \qquad N_{31} = 1$$
$$N_{14} = 7 \times 8 \times 6 = 336 \qquad N_{32} = 2$$
$$N_{15} = 4 \qquad N_{33} = 2$$
$$N_{16} = 3 \times 4 \times 4 = 48 \qquad N_{34} = 1$$
$$N_{17} = 2 \qquad N_{35} = 7$$
$$N_{18} = 2 \times 6 = 12 \qquad N_{36} = 7 \times 12 = 84 \qquad 5.32$$

As an example of the method of calculating the N_i, consider N_{13} and N_{14}. With a J_5 configuration, there are seven vacancy jumps that split the solute pair-vacancy complex; $N_{13} = 7$. There are also four equivalent vacancy sites at each end of the J_5 complex for each orientation of a solute pair. This gives eight configurations for each solute pair orientation. Consequently, there are $7 \times 8 = 56$ ways of forming a J_5 complex from a solute pair and a vacancy. Also, there are six distinguishable orientations of the solute pair for a total of $N_{14} = 7 \times 8 \times 6 = 336$.

From these reactions, one can obtain information on the rates of change of the variables defined: that is, dI_2/dt, dJ_6/dt, dJ_4/dt, dJ_2/dt, dJ_5/dt, dJ_3/dt, dJ_1/dt, dJ_0/dt, dC/dt, and the conservation of mass of the solute ($I_0 = I + C + 2J_0 + 2J_1 + 2J_2 + 2J_3 + 2J_4 + 2J_5 + 2J_6 + 2I_2$). If one adds to this list the rate of change of unassociated vacancies (dV/dt), 11 simultaneous equations result, and all the above variables can be found at equilibrium. That is, at equilibrium all the rates of change in these variables is zero, leaving 11 algebraic equations to solve.

This procedure is correct but cumbersome. Furthermore, the exact form of dV/dt is complicated by the interaction at sources and sinks. If one has other knowledge of V (the unassociated vacancy) all these other variables can be determined in terms of V. This gives the 10 equations listed as follows:

$$I_0 = I + C + 2(J_0 + J_1 + J_2 + J_3 + J_4 + J_5 + J_6 + I_2)$$

$$\frac{dC}{dt} = 0 = L_1 J_2 + L_3 J_4 + L_5 J_6 + L_{36} IV - (L_2 + L_4 + L_6) CI - L_{35} C$$

$$\frac{dJ_0}{dt} = 0 = -(L_7 + L_{21} + L_{33}) J_0 + L_8 I_2 V + L_{22} J_1 + L_{34} J_5$$

$$\frac{dJ_1}{dt} = 0 = -(L_9 + L_{22} + L_{23} + L_{25}) J_1 + L_{10} I_2 V + L_{24} J_2 + L_{26} J_5 + L_{21} J_0$$

$$\frac{dJ_3}{dt} = 0 = -(L_{11} + L_{27} + L_{29}) J_3 + L_{12} I_2 V + L_{28} J_4 + L_{30} J_5$$

$$\frac{dJ_5}{dt} = 0 = -(L_{13} + L_{26} + L_{30} + L_{31} + L_{34})J_5 + L_{14}I_2 V + L_{25}J_1$$
$$+ L_{29}J_3 + L_{32}J_6 + L_{33}J_0$$

$$\frac{dJ_2}{dt} = 0 = -(L_1 + L_{15} + L_{24})J_2 + L_2 CI + L_{16}I^2 V + L_{23}J_1$$

$$\frac{dJ_4}{dt} = 0 = -(L_3 + L_{17} + L_{28})J_4 + L_4 CI + L_{18}I^2 V + L_{27}J_3$$

$$\frac{dJ_6}{dt} = 0 = -(L_5 + L_{19} + L_{32})J_6 + L_6 CI + L_{20}I^2 V + L_{31}J_5$$

$$\frac{dI_2}{dt} = 0 = -(L_8 + L_{10} + L_{12} + L_{14})I_2 V + L_7 J_0 + L_9 J_1 + L_{11}J_3 + L_{13}J_5 \qquad 5.33$$

The V, however, will be nearly unchanged by these new variables; and with that approximation all the variables, J_i, and so forth, can be found in terms of V. Alternatively, one could express the free energy of the crystal in terms of all these variables. The reaction equations mentioned would then eliminate all independent variables but V from the free energy expression. Hence in principle V can be found exactly by minimizing the resultant expression. Each of those variables (J_i, etc.) are uniquely determined. From the form of Eqs. 5.33, it is obvious that these can be reduced to $J_i = \alpha_i CI + \beta_i I^2 V$, $i = 0, 1 \cdots 6$, and $I_2 = \gamma CI + \delta I^2$ where α_i, β_i, γ, and δ are constants. These results for J_i may be inserted into Eq. 5.33 to first eliminate C then solve the remaining cubic equation for I. All these variables would then be dependent only on V; and V can be determined independently.

The above procedure may be simplified by an approximation. The variables J_2, J_4, and J_6 are viewed as being nearly energetically equivalent. The same is true for J_1, J_3, J_5. Also, J_0 differs from J_1, J_3, or J_5 by a single solute-vacancy interaction. Suppose one represents the energetically similar configurations of these variables by a single variable. Let $J_2 + J_4 + J_6 = J$, and let $J_0 + J_1 + J_3 + J_5 = M$. The first 10 reactions in Eq. 5.30 reflect conversions of J and M to I, I_2, C and V. Reactions 11 through 17 are internal conversions of the new variables J and M. Because of the internal conversion reactions, it is not unreasonable to assume a free energy equivalence for these new variables. The free energy of J is the sum of the contributions from J_2, J_4, J_6. Since these configurations are energetically similar, the number of J configurations is defined as the sum of the contributions from the J_2, J_4, and J_6 configurations. Furthermore, if ϵ_2 is the fraction of total configurations in J that is supplied by J_2, then $J_2 = \epsilon_2 J$. Hence if the formation energies of J_2, J_4, and J_6 are equal, $J_2 = \frac{2}{7}J$, $J_4 = \frac{1}{7}J$, and $J_6 = \frac{4}{7}J$. With a minor modification due to the energy differences, the above reasoning leads to the free energy determining the distribution behavior J_2, J_4, J_6, and so forth. The reactions imply that not all of the jump frequencies are independent. The formation energy of J_2 should be independent of the two ways it can occur. This implies that $K_1/K_2 = K_{15}/K_{16}$, $K_3/K_4 = K_{17}/K_{18}$, and $K_5/K_6 = K_{17}/K_{18}$. Furthermore, this leads to the free energy of complex i, E_i, as $E_2 = \ln K_1/K_2$, $E_4 = \ln K_3/K_4$, $E_6 = \ln K_5/K_6$, $E_0 = \ln K_7/K_8$, $E_1 = \ln K_9/K_{10}$, $E_3 = \ln K_{11}/K_{12}$, and $E_5 = \ln K_{13}/K_{14}$. Thus using the free energy equivalence to determine the distribution,

$$J_2 = \frac{(2K_2/K_1)J}{2K_2/K_1 + K_4/K_3 + 4K_6/K_5}$$

$$J_4 = \frac{(K_4/K_3)J}{2K_2/K_1 + K_4/K_3 + 4K_6/K_5}$$

$$J_6 = \frac{(4K_6/K_5)J}{2K_2/K_1 + K_4/K_3 + 4K_6K_5}$$

$$J_0 = \frac{(4K_8/K_7)M}{4K_8/K_7 + 4K_{10}/K_9 + 2K_{12}/K_{11} + 8K_{14}/K_{13}}$$

$$J_1 = \frac{(4K_{10}/K_9)M}{4K_8/K_7 + 4K_{10}/K_9 + 2K_{12}/K_{11} + 8K_{14}/K_{13}}$$

$$J_3 = \frac{(2K_{12}/K_{11})M}{4K_8/K_7 + 4K_{10}/K_9 + 2K_{12}/K_{11} + 8K_{14}/K_{13}}$$

$$J_5 = \frac{(8K_{14}/K_{13})M}{4K_8/K_7 + 4K_{10}/K_9 + 2K_{12}/K_{11} + 8K_{14}/K_{13}} \qquad 5.34$$

Equilibrium for these new variables follows:

$$V \underset{L_2}{\overset{L_1}{\rightleftharpoons}} \text{sink source}$$

$$C \underset{L_4}{\overset{L_3}{\rightleftharpoons}} V + I$$

$$J \underset{L_6}{\overset{L_5}{\rightleftharpoons}} C + I$$

$$J \underset{L_8}{\overset{L_7}{\rightleftharpoons}} M$$

$$I_2 + V \underset{L_{10}}{\overset{L_9}{\rightleftharpoons}} M$$

$$J \underset{L_{12}}{\overset{L_{11}}{\rightleftharpoons}} V + 2I \qquad 5.35$$

The reaction rates for this reduced set of reactions is then the appropriate sum of those used in the reactions in Eq. 5.30. The following sums are appropriate.

$$\underline{L}_3 = L_{35}$$
$$\underline{L}_4 = L_{36}$$
$$\underline{L}_5 = L_1 + L_3 + L_5$$
$$\underline{L}_6 = L_2 + L_4 + L_6$$
$$\underline{L}_7 = L_{24} + L_{28} + L_{32}$$

$$\underline{L}_8 = L_{23} + L_{27} + L_{31}$$
$$\underline{L}_9 = L_8 + L_{10} + L_{12} + L_{14}$$
$$\underline{L}_{10} = L_7 + L_9 + L_{11} + L_{13}$$
$$\underline{L}_{11} = L_{15} + L_{17} + L_{19}$$
$$\underline{L}_{12} = L_{16} + L_{18} + L_{20}$$

The reactions shown in Eq. 5.35 lead to the following:

$$I_0 = I + C + 2(I_2 + J + M) \tag{5.36a}$$

$$\frac{dC}{dt} = 0 = -\underline{L}_3 C + \underline{L}_4 VI + \underline{L}_5 J - \underline{L}_6 CI \tag{5.36b}$$

$$\frac{dJ}{dt} = 0 = -(\underline{L}_5 + \underline{L}_7 + \underline{L}_{11})J + \underline{L}_6 CI + \underline{L}_8 M + \underline{L}_{12} I^2 V \tag{5.36c}$$

$$\frac{dM}{dt} = 0 = -(\underline{L}_{10} + \underline{L}_8)M + \underline{L}_7 J + \underline{L}_9 I_2 V \tag{5.36d}$$

$$\frac{dI_2}{dt} = 0 = \underline{L}_{10} M - \underline{L}_9 I_2 V \tag{5.36e}$$

A partial solution to this set yields

$$J = \frac{\underline{L}_6 CI + \underline{L}_{12} I^2 V}{\underline{L}_5 + \underline{L}_{11}}$$

$$M = \frac{\underline{L}_7}{\underline{L}_8} J$$

and

$$I_2 = \frac{\underline{L}_{10}\underline{L}_7}{\underline{L}_9\underline{L}_8} \frac{J}{V}$$

Hence one can write

$$J = \alpha_J CI + \beta_J I^2 V$$
$$M = \alpha_M CI + \beta_M I^2 V$$

and

$$I_2 = \alpha_I CI + \beta_I I^2 V$$

Using these relations in Eq. 5.36b–e, one obtains

$$C = \frac{\underline{L}_4 VI + \underline{L}_5 \beta_J I^2 V}{\underline{L}_3 + \underline{L}_6 I - \underline{L}_5 \alpha_J I}$$

Finally, this can be substituted into Eq. 5.36a to give a cubic equation for the determination of I dependent only on V. Hence

$$I^3\{(\beta_I+\beta_J+\beta_M)(L_6-L_5\alpha_J)2V+(\alpha_I+\alpha_J+\alpha_M)2L_5\beta_J V\}$$
$$+I^2\{(\beta_I+\beta_J+\beta_M)2L_3 V+L_6-L_5\alpha_J+L_5\beta_J V+2(\alpha_I+\alpha_J+\alpha_M)L_4 V\}$$
$$+I\{L_3-(L_6-\alpha_J L_5)I_0+L_4 V\}-L_3 I_0=0=AI^3+BI^2+EI+D=0. \quad 5.37$$

On solving the cubic equation, 5.37, all the variables I_2, I, J, M, and C are found to be dependent on V; and V is assumed to be determined separately. With these results, the factors J_i can be found using Eq. 5.34.

All solute tracer jumps influence the flux; one can, therefore, express the diffusion coefficient in terms of the variables introduced to account for the configuration in which the tracer is found. The diffusion coefficient can be calculated as given in Eq. 5.16,

$$D=\frac{b^2}{3}\langle W\rangle f$$

The average jump frequency can be directly obtained from the configurational results by Eq. 5.14.

$$\langle W\rangle = w_2\frac{C}{I_0}+\frac{J_0 K_0}{I_0}+\frac{J_1 K_{23}}{I_0}+\frac{J_3 K_{27}}{I_0}+\frac{J_5 K_{31}}{I_0}$$
$$+\frac{J_2 K_{24}}{I_0}+\frac{J_4 K_{28}}{I_0}+\frac{J_6 K_{32}}{I_0} \quad 5.38$$

The calculation of the correlation factor follows Eq. 5.13. Let the direction for diffusion in the fcc lattice be parallel to the [100] direction. For the types of jumps taken by the solute tracer atom, the f_α are the partial correlation factors, and the C_α are the relative probabilities of occurrence. The different jumps that appear in Eq. 5.12 are all related to the solute jumps that were described previously; the eight different jump types appearing in $\langle W\rangle$ must be further subdivided for the calculation of f_α. This is necessary to account for different orientations of the reaction products relative to the X axis. Table 5.2 identifies the different jump types and the frequency of occurrence. In Table 5.2 the solute tracer is initially at the origin, with the vacant lattice site at $b(1,1,0)$. The vacancy-tracer jump is into the origin.

The calculation of the partial correlation factors follows from the determination of the 13×13 element matrix; P_2 is given by Eqs. 5.27, 5.28, and 5.29.

Table 5.2 Definition of the 13 types of tracer jumps[a]

Type (α)	Second impurity position	Tracer jump frequency	Configuration variable	C_α
1	$b(1, 0, 1)$	K_0	J_0	$J_0 K_0/2\gamma$[b]
2	$b(0, 1, 1)$	K_0	J_0	$J_0 K_0/2\gamma$
3	$b(2, 0, 0)$	K_{24}	J_2	$J_2 K_{24}/2\gamma$
4	$b(2, 1, 1)$	K_{32}	J_6	$J_6 K_{32}/2\gamma$
5	$b(2, 2, 0)$	K_{28}	J_4	$J_4 K_{28}/\gamma$
6	$b(1, 2, 1)$	K_{32}	J_6	$J_6 K_{32}/2\gamma$
7	$b(0, 2, 0)$	K_{24}	J_2	$J_2 K_{24}/2\gamma$
8	$b(1, -1, 0)$	K_{23}	J_1	$J_1 K_{23}/2\gamma$
9	$b(0, -1, 1)$	K_{31}	J_5	$J_5 K_{31}/2\gamma$
10	$b(-1, -1, 0)$	K_{27}	J_3	$J_3 K_{27}/\gamma$
11	$b(-1, 0, 1)$	K_{31}	J_5	$J_5 K_{31}/2\gamma$
12	$b(-1, 1, 0)$	K_{23}	J_1	$J_1 K_{23}/2\gamma$
13	Not present	w_2	C	Cw_2/γ

[a] J. P. Stark, *J. Appl. Phys.* **43**, 4404 (1972).
[b] $\gamma = Cw_2 + J_0 K_0 + J_1 K_{23} + J_3 K_{27} + J_5 K_{31} + J_2 K_{24} + J_4 K_{28} + J_6 K_{32}$

One can group together sets of configurations that have n-fold rotational or mirror symmetry. With this reduction in the number of groups of equivalent configurations, one can reduce this set to 21 elements where the tracer, solute, or vacancy is a nearest neighbor of the other two, as in the previous problem. If one were to take the set of sites one vacancy jump removed from these, 98 additional sets would result. This set is too large for normal computation. Consequently, only the 21 sets are to be used herein. One element is added to account for the correlation factor at infinite dilution; the vacancy jumps around an isolated solute are thus separated from the others. With this addition, the matrix A is a 22×22 element matrix.

The matrices A and P_2 are not generally symmetric. Consequently, care must be used in the evaluation of Eq. 5.29. One may order the vectors R and Q into rectangular matrices. With this ordering, R can be written as a matrix S with 22 rows and 13 columns. The Q becomes a matrix U with 13 rows and 22 columns. With V as the transpose of P_2, one can write the following:

$$P_2^T = V = U(1 - A)^{-1} S \qquad\qquad 5.39$$

That is, the elements $V_{\alpha\beta} = P_2^{\beta\alpha}$. Also, Table 5.3 identifies the configurations in the matrices.

Table 5.3 Configuration classes[a]

Configuration in A_{ij}	Tracer $(0, 0, 0)$		Number of configurations per class	Jump configuration (α)
	Solute	Vacancy		
1	$(\bar{1}, 1, 0)$	$(1, 1, 0)$	4	12
2	$(1, 1, 0)$	$(\bar{1}, \bar{1}, 0)$	4	10
3	$(\bar{1}, 0, 1)$	$(1, 1, 0)$	8	11
4	$(2, 0, 0)$	$(1, \bar{1}, 0)$	4	3
5	$(2, 2, 0)$	$(1, 1, 0)$	4	5
6	$(2, 1, 1)$	$(1, 1, 0)$	8	4
7	$(1, \bar{1}, 0)$	$(2, 0, 0)$	4	—
8	$(1, 1, 0)$	$(2, 2, 0)$	4	—
9	$(1, 0, 1)$	$(2, 1, 1)$	8	—
10	$(0, \bar{1}, 1)$	$(1, 0, 1)$	8	2
11	$(0, 2, 0)$	$(1, 1, 0)$	4	7
12	$(0, \bar{1}, 1)$	$(1, 1, 0)$	8	9
13	$(0, 1, 1)$	$(1, 2, 1)$	8	—
14	$(1, 0, 1)$	$(1, 1, 0)$	8	1
15	$(1, \bar{1}, 0)$	$(1, 1, 0)$	4	8
16	$(1, 2, 1)$	$(1, 1, 0)$	8	6
17	$(1, 1, 0)$	$(1, 2, 1)$	8	—
18	$(1, 1, 0)$	$(0, 1, 1)$	8	—
19	$(1, \bar{1}, 0)$	$(0, 1, 1)$	8	—
20	$(1, 1, 0)$	$(0, 2, 0)$	4	—
21	$(1, 2, 1)$	$(0, 1, 1)$	8	—

[a] J. P. Stark, ibid.

To facilitate writing the matrix elements, it is convenient to let

$$\alpha_1 = 2K_0 + K_{37} + 2K_{21} + 2K_{33} + 5K_7$$

$$\alpha_2 = 2K_{40} + 2K_{24} + 4K_{15} + 4K_1$$

$$\alpha_3 = K_{38} + 2K_{22} + 2K_{25} + 6K_9 + K_{23}$$

$$\alpha_4 = 2K_{28} + 2K_{17} + 8K_3$$

$$\alpha_5 = K_{27} + 4K_{29} + 7K_{11}$$

$$\alpha_6 = 2K_{32} + K_{41} + 5K_5 + 4K_{19}$$

$$\alpha_7 = K_{31} + 7K_{13} + K_{26} + K_{30} + K_{39} + K_{34}$$

$$\alpha_8 = 2w_1 + w_2 + 7Fw_3$$

The F is the probability of the vacancy leaving the isolated solute, and is the usual function of w_4/w_0.

Consistent with the above, the nonzero matrix elements of S, U, and A are presented as

$$S_{10,1} = \tfrac{1}{8} \qquad\qquad S_{11,8} = \tfrac{1}{4}$$
$$S_{14,2} = \tfrac{1}{8} \qquad\qquad S_{16,9} = \tfrac{1}{8}$$
$$S_{1,3} = \tfrac{1}{4} \qquad\qquad S_{5,10} = \tfrac{1}{4}$$
$$S_{3,4} = \tfrac{1}{8} \qquad\qquad S_{6,11} = \tfrac{1}{8}$$
$$S_{2,5} = \tfrac{1}{4} \qquad\qquad S_{4,12} = \tfrac{1}{4}$$
$$S_{12,6} = \tfrac{1}{8} \qquad\qquad S_{22,13} = 1$$
$$S_{15,7} = \tfrac{1}{4}$$

$$U_{1,14} = +8K_0/\alpha_1 \qquad\qquad U_{2,10} = +8K_0/\alpha_1$$
$$U_{3,4} = +4K_{24}/\alpha_2 \qquad\qquad U_{4,6} = +8K_{32}/\alpha_6$$
$$U_{5,5} = +4K_{28}/\alpha_4 \qquad\qquad U_{6,6} = +8K_{32}/\alpha_6$$
$$U_{7,11} = +4K_{24}/\alpha_2 \qquad\qquad U_{8,15} = +4K_{23}/\alpha_3$$
$$U_{9,12} = +8K_{31}/\alpha_7 \qquad\qquad U_{10,2} = +4K_{27}/\alpha_5$$
$$U_{11,3} = +8K_{31}/\alpha_7 \qquad\qquad U_{12,1} = +4K_{23}/\alpha_3$$
$$U_{13,22} = +w_2/\alpha_8$$

$$A_{20,1} = K_{38}/\alpha_3 \qquad\qquad A_{18,1} = K_{22}/\alpha_3$$
$$A_{3,1} = K_{25}/\alpha_3 \qquad\qquad A_{19,2} = K_{29}/\alpha_5$$
$$A_{3,2} = K_{29}/\alpha_5 \qquad\qquad A_{1,3} = 2K_{26}/\alpha_7$$
$$A_{2,3} = 2K_{30}/\alpha_7 \qquad\qquad A_{18,3} = K_{34}/\alpha_7$$
$$A_{19,3} = K_{39}/\alpha_7 \qquad\qquad A_{4,4} = 2K_{40}/\alpha_7$$
$$A_{7,4} = K_{24}/\alpha_7 \qquad\qquad A_{8,5} = K_{28}/\alpha_4$$
$$A_{9,6} = K_{32}/\alpha_6 \qquad\qquad A_{6,6} = K_{41}/\alpha_6$$
$$A_{15,7} = K_{38}/\alpha_3 \qquad\qquad A_{14,7} = K_{22}/\alpha_3$$
$$A_{9,7} = K_{25}/\alpha_3 \qquad\qquad A_{4,7} = K_{23}/\alpha_3$$
$$A_{9,8} = K_{29}/\alpha_5 \qquad\qquad A_{17,8} = K_{29}/\alpha_5$$
$$A_{5,8} = K_{27}/\alpha_5 \qquad\qquad A_{6,9} = K_{31}/\alpha_7$$
$$A_{7,9} = 2K_{26}/\alpha_7 \qquad\qquad A_{8,9} = 2K_{30}/\alpha_7$$
$$A_{14,9} = K_{34}/\alpha_7 \qquad\qquad A_{17,9} = K_{39}/\alpha_7$$
$$A_{18,10} = K_0/\alpha_1 \qquad\qquad A_{10,10} = K_{37}/\alpha_1$$
$$A_{12,10} = K_{33}/\alpha_1 \qquad\qquad A_{13,10} = K_{33}/\alpha_1$$
$$A_{20,11} = K_{24}/\alpha_2 \qquad\qquad A_{10,12} = K_{34}/\alpha_7$$

$$A_{12,12} = K_{39}/\alpha_7 \qquad A_{21,13} = K_{31}/\alpha_7$$
$$A_{10,12} = K_{34}/\alpha_7 \qquad A_{13,13} = K_{39}/\alpha_7$$
$$A_{14,14} = K_0/\alpha_1 \qquad A_{18,14} = K_{37}/\alpha_1$$
$$A_{15,14} = 2K_{21}/\alpha_1 \qquad A_{7,14} = 2K_{21}/\alpha_1$$
$$A_{19,14} = K_{33}/\alpha_1 \qquad A_{9,14} = K_{33}/\alpha_1$$
$$A_{7,15} = K_{38}/\alpha_3 \qquad A_{14,15} = K_{22}/\alpha_3$$
$$A_{19,15} = K_{25}/\alpha_3 \qquad A_{19,16} = K_{32}/\alpha_6$$
$$A_{21,16} = K_{41}/\alpha_6 \qquad A_{16,17} = K_{31}/\alpha_7$$
$$A_{20,17} = 2K_{26}/\alpha_7 \qquad A_{18,17} = K_{34}/\alpha_7$$
$$A_{9,17} = K_{39}/\alpha_7 \qquad A_{8,17} = 2K_{30}/\alpha_7$$
$$A_{10,18} = K_0/\alpha_1 \qquad A_{14,18} = K_{37}/\alpha_1$$
$$A_{1,18} = 2K_{21}/\alpha_1 \qquad A_{20,18} = 2K_{21}/\alpha_1$$
$$A_{3,18} = K_{33}/\alpha_1 \qquad A_{17,18} = K_{33}/\alpha_1$$
$$A_{18,18} = K_0/\alpha_1 \qquad A_{2,19} = 2K_{30}/\alpha_7$$
$$A_{15,19} = 2K_{26}/\alpha_7 \qquad A_{3,19} = K_{39}/\alpha_7$$
$$A_{14,19} = K_{34}/\alpha_7 \qquad A_{21,19} = K_{31}/\alpha_7$$
$$A_{1,20} = K_{38}/\alpha_3 \qquad A_{18,20} = 2K_{22}/\alpha_3$$
$$A_{17,20} = 2K_{25}/\alpha_3 \qquad A_{11,20} = K_{23}/\alpha_3$$
$$A_{13,21} = K_{32}/\alpha_6 \qquad A_{16,21} = K_{41}/\alpha_6$$
$$A_{19,21} = K_{32}/\alpha_6$$

To illustrate the variation in parameters resulting from the calculation, some simplifications are necessary. In particular, we show the importance of two parameters in this theory. The binding energy between a solute pair and a solute and vacancy is investigated numerically. In this regard, the calculations reflect the bond energy affecting only the breaking up of a pair and not the forming of a pair. Such an approximation is an over-simplification, but may provide a useful guide to the determination of the jump frequencies K_i.

Therefore, let

$$K_0 = Zw_2 \qquad K_{19} = w_3 e^{-E_B}$$
$$K_1 = w_3 \qquad K_{20} = w_4$$
$$K_2 = w_4 \qquad K_{21} = w_1 e^{-E_B}$$
$$K_3 = w_3 \qquad K_{22} = w_1$$

$$K_4 = w_4 \qquad\qquad K_{23} = Zw_2 e^{-E}$$

$$K_5 = w_3 \qquad\qquad K_{24} = Zw_2 e^{-E_B}$$

$$K_6 = w_4 \qquad\qquad K_{25} = w_1$$

$$K_7 = w_3 e^{-E_B} \qquad K_{26} = w_1$$

$$K_8 = w_4 \qquad\qquad K_{27} = Zw_2 e^{-E}$$

$$K_9 = w_3 \qquad\qquad K_{28} = Zw_2 e^{-E_B}$$

$$K_{10} = w_4 \qquad\qquad K_{29} = w_1$$

$$K_{11} = w_3 \qquad\qquad K_{30} = w_1$$

$$K_{12} = w_4 \qquad\qquad K_{31} = Zw_2 e^{-E}$$

$$K_{13} = w_3 \qquad\qquad K_{32} = Zw_2 e^{-E_B}$$

$$K_{14} = w_4 \qquad\qquad K_{33} = w_1 e^{-E_B}$$

$$K_{15} = w_3 e^{-E_B} \qquad K_{34} = w_4$$

$$K_{16} = w_4 \qquad\qquad K_{35} = w_3$$

$$K_{17} = w_3 e^{-E_B} \qquad K_{36} = w_4$$

$$K_{18} = w_4 \qquad\qquad K_{37} = K_{38} = K_{39} = K_{40} = K_{41} = w_1$$

The Z influences the solute jump rate in the vicinity of the complexes; E and E_B represent the appropriate energy term divided by $K_B T$.

A further simplification is introduced so that the independent variables are Z, E, and E_B; namely, let $w_1 = w_2 = w_4 = w_3 e^{E_B}$. With the above, Table 5.4 gives a list of the partial correlation factors consistent with these approximations.

The values in Table 5.4 give an indication of the degree that this calculation approximates the correlation factor. First of all, the values for $\alpha = 13$ are exact. When $Z = e^{-E_B} = e^{-E} = 1$, the partial correlation factors for cases 1 through 12 are expected to equal $f_{13} = 0.78$. The error is about 6.5%. Another indication of the accuracy is the situation in which $Z \to \infty$ and $e^{-E_B} = e^{-E} \to 0$. An independent and exact calculation for this case was possible for f_1 and f_2. These were found to be $f_1 = +0.0$ and $f_2 = -0.0$. When $Z = 100$, $e^{-E_B} = e^{-E} = 0.01$, then $f_1 = +0.01$ and $f_2 = -0.01$; also, when $Z = 10$, $e^{-E_B} = e^{-E} = 0.1$, then $f_1 = 0.08$ and $f_2 = -0.05$, as found in Table 5.4.

Table 5.5 presents comparative calculations of jump frequency and the probability of a solute vacancy pair. One can calculate the probability of a solute tracer having a neighboring vacancy for the case of infinite dilution as $12V \exp(E_B/KT) = C$. This value should compare to calculations from Eq. 5.37 in which one evaluates I, C, J, and M. The comparison is

Table 5.4 Partial correlation factors

e^{-E_B}	e^{-E}	f_1	f_2	f_3	f_4	f_5	f_6	f_7	f_8	f_9	f_{10}	f_{11}	f_{12}	f_{13}
						For $Z=1$								
1.0	1.0	0.81	0.80	0.76	0.81	0.76	0.81	0.75	0.84	0.84	0.84	0.83	0.81	0.78
0.1	1.0	0.20	0.10	-0.35	0.23	-0.26	0.44	-0.32	0.94	0.81	0.87	0.78	0.74	0.55
1.0	0.1	0.82	0.82	0.96	0.97	0.96	0.96	0.90	0.83	0.83	0.83	0.81	0.79	0.78
0.1	0.1	0.21	0.09	-0.13	0.48	0.09	0.74	-0.25	0.93	0.77	0.81	0.73	0.67	0.55
						For $Z=10$								
1.0	1.0	0.40	0.21	0.07	0.23	0.10	0.22	0.07	0.50	0.48	0.51	0.46	0.42	0.78
0.1	1.0	0.05	-0.02	-0.54	-0.23	-0.64	-0.19	-0.58	0.77	0.54	0.71	0.51	0.47	0.55
1.0	0.1	0.45	0.22	0.70	0.83	0.79	0.83	0.55	0.40	0.35	0.35	0.31	0.24	0.78
0.1	0.1	0.08	-0.05	-0.33	0.29	-0.17	0.53	-0.46	0.61	0.27	0.36	0.24	0.16	0.55

[a] J. P. Stark, ibid.

Table 5.5 Probability of solute vacancy pair[ab]

E_B (kcal)	E (kcal)	Solute (%)	T (°K)	$Z=1$ $12\,Ve^{E_B}$	$\dfrac{C+J+M}{I_0}$	$\dfrac{\langle W(I_0)\rangle}{\langle W(I_0\to 0)\rangle}$
0.0	0.0	1.00	700	7.92×10^{-7}	7.97×10^{-7}	1.00
		0.10			7.91×10^{-7}	1.00
		0.01			7.89×10^{-7}	1.00
		1.00	1000	1.125×10^{-4}	1.14×10^{-4}	1.02
		0.10			1.13×10^{-4}	1.01
		0.01			1.13×10^{-4}	1.00
2.0	0.0	1.00	700	3.34×10^{-6}	4.32×10^{-6}	1.03
		0.10			3.47×10^{-6}	1.00
		0.01			3.34×10^{-6}	1.00
		1.00	1000	3.08×10^{-4}	3.59×10^{-4}	1.03
		0.10			3.15×10^{-4}	1.00
		0.01			3.08×10^{-4}	1.00
0.0	2.0	1.00	700	7.92×10^{-7}	8.20×10^{-7}	1.04
		0.10			7.95×10^{-7}	0.99
		0.01			7.89×10^{-7}	1.00
		1.00	1000	1.125×10^{-4}	1.16×10^{-4}	1.03
		0.10			1.12×10^{-4}	1.00
		0.01			1.12×10^{-4}	1.00
2.0	2.0	1.00	700	3.34×10^{-6}	4.32×10^{-6}	1.10
		0.10			3.51×10^{-6}	1.02
		0.01			3.34×10^{-6}	1.00
		1.00	1000	3.08×10^{-4}	3.06×10^{-4}	1.06
		0.10			3.17×10^{-4}	1.01
		0.01			3.08×10^{-4}	1.00
10.0	0.0	1.00	700	1.04×10^{-3}	8.97×10^{-2}	0.855
		0.10			1.73×10^{-2}	0.98
		0.01			2.86×10^{-3}	1.00
		1.00	1000	1.73×10^{-2}	1.46×10^{-1}	0.735
		0.10			4.43×10^{-2}	1.02
		0.01			1.71×10^{-2}	1.00
0.0	10.0	1.00	700	7.92×10^{-7}	8.93×10^{-7}	1.13
		0.10			8.78×10^{-7}	1.11
		0.01			8.46×10^{-7}	1.07
		1.00	1000	1.125×10^{-4}	1.25×10^{-4}	1.11
		0.10			1.21×10^{-4}	1.08
		0.01			1.15×10^{-4}	1.02
10.0	10.0	1.00	700	1.04×10^{-3}	2.15×10^{-3}	1.35
		0.10			2.01×10^{-3}	1.35
		0.01			1.68×10^{-3}	1.23

Table 5.5 (*continued*)

E_B (kcal)	E (kcal)	Solute (%)	T (°K)	$Z = 1$ $12\,Ve^{E_B}$	$\dfrac{C + J + M}{I_0}$	$\dfrac{\langle W(I_0)\rangle}{\langle W(I_0 \to 0)\rangle}$
		1.00	1000	1.73×10^{-2}	3.13×10^{-2}	1.29
		0.10			2.68×10^{-2}	1.20
		0.01			2.03×10^{-2}	1.05
				$Z = 10$		
10.0	0.0	1.00	700	1.04×10^{-3}	8.97×10^{-2}	2.6
		0.10			1.73×10^{-2}	1.3
		0.01			2.86×10^{-3}	1.0
0.0	10.0	1.00	700	7.92×10^{-7}	8.93×10^{-7}	10.7
		0.10			8.78×10^{-7}	9.5
		0.01			8.46×10^{-7}	6.5
10.0	10.0	1.00	700	1.04×10^{-3}	2.15×10^{-3}	12.9
		0.10			2.01×10^{-3}	10.6
		0.01			1.68×10^{-3}	6.2

[a] J. P. Stark, ibid.
[b] Values calculated are based on $\nu = 5 \times 10^{12}\,\text{sec}^{-1}$, $V = \exp -(1\,eV/KT)$, $w_0 = w_1 = w_4 = w_3 \exp(E_B/KT) = \nu \exp -(1\,eV/KT)$.

$(C + J + M)/I_0$ being the probability of a solute tracer having a neighboring vacancy. As can be seen in the table this probability can depart from the infinite dilution case significantly. Table 5.5 also presents values of the ratio of the jump frequency for the concentration given, $\langle W(I_0)\rangle$, to infinite dilution $\langle W(I_0 \to 0)\rangle$. When $Z = 1$, one can see that the increased probability of having a neighboring vacancy does not influence the jump rate appreciably. That is, for $E_B = 10.0$ and $E = 0$, the solute in the 1% alloy has a vacancy nearly 90 times more frequently at 700°K than at infinite dilution. The jump rate, however, has decreased slightly. The decrease must be associated with the increased stability of the complex so formed. Since a large vacancy concentration around a solute does not influence the solute jump rate, it was of interest to see the influence of a change in the vacancy-tracer exchange frequency. Consequently the values for $Z = 10$ were introduced in the program to increase vacancy-solute exchanges in the solute-solute (tracer)-vacancy complex. It is notable that a strong solute vacancy bond is counteracted by a large solute pair energy. Indeed, it appears that one of the only ways that solute pairs could increase the diffusivity is through a large solute pair bonding energy and a large jump rate of vacancies into such pairs. In all other situations, the tendency is for the diffusivity to be decreased by the formation of solute pairs. The correlation factor tends to offset any increased jump rate. A substantial decrease in the jump rate, however, would not be offset by a corresponding correlation factor increase. The net result would be a decreased diffusivity for a decreased jump rate of a vacancy in the region of a solute pair.

INFLUENCE OF VACANCY PAIRS ON THE SOLUTE DIFFUSIVITY

In most treatments of vacancy pairs that influence the diffusivity of a substitutional solute in an fcc crystal, the authors assume the vacancies are bound with an infinite energy so that they form a divacancy. Such a treatment is an oversimplification that may give misleading results. To avoid this assumption, one must treat a large number of "chemical" reactions similar to those given in Eq. 5.30. The discussion here parallels that of McKee and Stark.[6]

One defines the reaction complexes as the following:

I_0 atom fraction of solute atoms.

$I =$ fraction of lattice sites occupied by unassociated solute atoms.

$C =$ fraction of lattice sites occupied by solute atoms with an associated single vacancy.

$V =$ fraction of lattice sites occupied by single vacancies.

$V_2 =$ fraction of lattice sites occupied by vacancy pairs.

$P_0 =$ fraction of lattice sites occupied by solute atoms with two associated vacancies at 60°.

$P_1 =$ fraction of lattice sites occupied by vacancy solute pairs with an associated vacancy at 90°.

P_2 fraction of lattice sites occupied by divacancies with an associated solute atom at 90°.

$P_3 =$ fraction of lattice sites occupied by vacancy-solute pairs with an associated vacancy at 180°.

P_4 fraction of lattice sites occupied by divacancies with an associated solute at 180°.

P_5 fraction of lattice site occupied by vacancy-solute pairs with an associated single vacancy at 120°.

$P_6 =$ fraction of lattice sites occupied by divacancies with an associated solute at 120°.

Also, it will be necessary to introduce

$$J = P_2 + P_4 + P_6$$

and

$$M = P_0 + P_1 + P_3 + P_5$$

The method of solution follows; 4 parallels the previous section. One can write 34 reversible reactions between the configurations I, C, V, V_2, P_i, and I_0, based on a single vacancy jump. From these one can deduce a set of kinetic equations similar to Eq. 5.33. An approximate solution may be found by reducing the P_i variable to the J and M, and writing a set of reduced reactions similar to Eq. 5.35. The reduced reactions give a new but easier set of reduced kinetic equations to Eq. 5.36. The reduced kinetic equations can be solved to yield a quadratic equation to Eq. 5.37. The net result is a very messy but analytical one for the P_i's. If the tracer-vacancy exchange rate is K_i for complex P_i, then it is apparent that the average frequency of the tracer may be written

$$\langle W \rangle = w_2 \frac{C}{I_0} + \sum_{i=0}^{6} \frac{K_i P_i}{I_0} \qquad 5.40$$

where C and P_i are determined as above. The correlation factor is also needed. With the x axis perpendicular to (100) planes, the jump configurations that are related by N-fold rotational symmetry around the x axis are defined in Table 5.6, with the tracer at the origin and the two vacancy

Table 5.6 Divacancy correlation complexes[a]

Tracer at 0, 0, 0					
Vacant site 1	Vacant site 2	Jump frequency	Decomposition variable	Type α	C_α
$(1,1,0)$	$(-1,1,0)$	K_1	P_1	1	$K_1 P_1 / 2\gamma$[b]
$(1,1,0)$	$(-1,-1,0)$	K_3	P_3	2	$K_3 P_3 / \gamma$
$(1,1,0)$	$(-1,0,-1)$	K_5	P_5	3	$K_5 P_5 / 2\gamma$
$(1,0,-1)$	$(0,-1,-1)$	K_0	P_0	4	$K_0 P_0 / 2\gamma$
$(1,1,0)$	$(0,-1,-1)$	K_5	P_5	5	$K_5 P_5 / 2\gamma$
$(1,0,-1)$	$(0,0,-2)$	K_2	P_2	6	$K_2 P_2 / 2\gamma$
$(1,1,0)$	$(1,0,-1)$	K_0	P_0	7	$K_0 P_0 / 2\gamma$
$(1,1,0)$	$(1,-1,0)$	K_1	P_1	8	$K_1 P_1 / 2\gamma$
$(1,1,0)$	$(1,2,1)$	K_0	P_0	9	$K_0 P_0 / 2\gamma$
$(1,1,0)$	$(2,0,0)$	K_2	P_2	10	$K_2 P_2 / 2\gamma$
$(1,1,0)$	$(2,2,0)$	K_4	P_4	11	$K_4 P_4 / \gamma$
$(1,1,0)$	$(2,1,-1)$	K_6	P_6	12	$K_6 P_6 / 2\gamma$
$(1,1,0)$	Not present	w_2	C	13	$C w_2 / \gamma$

[a] R. A. McKee and J. P. Stark, *Phys. Rev.* **B7**, 613 (1973).

[b] $\gamma = C w_2 + \sum_{i=0}^{6} K_i P_i$.

positions indicated. The C_α are also defined in that table. One therefore needs only the partial correlation factors to determine the diffusivity; it is presumed that computer calculations have determined the P_i and C, as previously outlined.

It would now seem that Eq. 5.29 should be used to aid in calculating the partial correlation factors. Unfortunately the reduced matrix A indicated therein presents a problem. The variables P_1, P_3, and P_5 can form a mirror image of themselves. These three complexes are formed by a vacancy jump from either P_2, P_4, or P_6, or a mirror image of P_2, P_4, or P_6. Hence one will not know in the A matrix used in Eq. 5.29 whether or not the previous jump is from P_2 or its mirror image. The same is true for P_4 and P_6. This knowledge is necessary in the specification of $P_{\alpha\beta}^{11}$ or $P_{\alpha\beta}^a$ using $R(\alpha)$ or $R(\alpha^I)$.

As a consequence of the additional symmetry introduced by P_1, P_3, and P_5, one must use the calculation of $P_{\alpha\beta}^a$ as indicated in Eq. 5.28, where the A matrix distinguishes between complex P_2 and its mirror image through the plane containing the tracer and perpendicular to the x axis. As a consequence, the A matrix must contain the following: three configurations where the image is the same configuration, nine configurations plus their mirror images where the images are distinguishable, and one configuration and its image where neither vacancy can cause a jump in the x direction. This gives a total of 23 configurations contained in the A matrix. Now since one is only interested in the $P_2 = P_{\alpha\beta}^{11} - P_{\alpha\beta}^a$, those configurations where both vacancies and the tracer lie on the same x plane can be omitted. Such configurations would contribute equally to both terms.

With the 34 reactions leading to the set of configurations P_i, one now has a total of 62 vacancy jump frequencies, K_i. These may be reduced by a simplifying model so that computer calculations of the diffusivity are possible. To accomplish this, McKee and Stark[7] let E_B be the binding energy between the tracer and a vacancy so that $X = e^{-E_B/K_B T}$. The E_V is the binding energy between two vacancies so that $Y = e^{-E_V/K_B T}$. They also define Z_1 to be factor, to allow an increased jump rate of solvent atoms into divacancies and Z_2 for solute atoms into the divacancy. Based on such a model with four variables, including the solute composition I_0, they tabulate the diffusivity. These tabulations are given in Table 5.7. In addition, they consider self-diffusion as influenced by this model. Table 5.8 gives their values for that calculation.

An indication of the accuracy of the calculation technique can be seen in the partial correlation factor for the P_0 complex. If this complex is formed and there is an infinite binding between vacancies and between a solute and a vacancy, the correlation factor for this bound complex will go

Table 5.7 Solute diffusion[a]

f_T	f_C	$\langle W \rangle f_T$	X	Y
$I_0 = 0.1\%$; $Z_2 = 1$; $Z_1 = 1$				
0.781	0.781	4.139×10^3	1.00	1.00
0.557	0.557	2.949×10^4	0.10	1.00
0.781	0.781	4.137×10^3	1.00	0.10
0.556	0.557	2.945×10^4	0.10	0.10
0.506	0.506	2.655×10^5	0.01	1.00
0.494	0.506	2.631×10^5	0.01	0.01
$I_0 = 10\%$; $Z_2 = 1$; $Z_1 = 1$				
0.781	0.781	4.139×10^3	1.00	1.00
0.554	0.557	2.949×10^4	0.10	1.00
0.781	0.781	4.140×10^3	1.00	0.10
0.556	0.557	2.949×10^4	0.10	0.10
0.506	0.506	2.657×10^5	0.01	1.00
0.494	0.506	2.653×10^5	0.01	0.01
$I_0 = 0.1\%$; $Z_2 = 100$; $Z_1 = 1$				
0.774	0.781	4.153×10^3	1.00	1.00
0.510	0.557	2.969×10^4	0.10	1.00
0.763	0.781	4.148×10^3	1.00	0.10
0.441	0.557	2.964×10^4	0.10	0.10
0.289	0.506	2.744×10^5	0.01	1.00
0.133	0.506	2.677×10^5	0.01	0.01
$I_0 = 10\%$; $Z_2 = 100$; $Z_1 = 1$				
0.774	0.781	4.153×10^3	1.00	1.00
0.510	0.557	2.969×10^4	0.10	1.00
0.761	0.781	4.151×10^3	1.00	0.10
0.439	0.557	2.966×10^4	0.10	0.10
0.289	0.506	2.744×10^5	0.01	1.00
0.130	0.506	2.691×10^5	0.01	0.01
$I_0 = 0.1\%$; $Z_2 = 1$; $Z_1 = 100$				
0.781	0.781	4.137×10^3	1.00	1.00
0.557	0.557	2.945×10^4	0.10	1.00
0.781	0.781	4.133×10^3	1.00	0.10
0.557	0.557	2.935×10^4	0.10	0.10
0.506	0.506	2.651×10^5	0.01	1.00
0.508	0.506	2.661×10^5	0.01	0.01
$I_0 = 10\%$; $Z_2 = 1$; $Z_1 = 100$				
0.781	0.781	4.137×10^3	1.00	1.00
0.557	0.557	2.946×10^4	0.10	1.00
0.781	0.781	4.137×10^3	1.00	0.10

(continued overleaf)

Table 5.7 (*continued*)

f_T	f_C	$\langle W \rangle f_T$	X	Y
0.557	0.557	2.948×10^4	0.10	0.10
0.508	0.506	2.662×10^5	0.01	1.00
0.508	0.506	2.701×10^5	0.01	0.01
$I_0 = 0.1\%$; $Z_2 = 100$; $Z_1 = 100$				
0.781	0.781	4.139×10^3	1.00	1.00
0.557	0.557	2.954×10^4	0.10	1.00
0.781	0.781	4.139×10^3	1.00	0.10
0.546	0.557	2.963×10^4	0.10	0.10
0.506	0.506	2.651×10^5	0.01	1.00
0.215	0.506	2.737×10^5	0.01	0.01
$I_0 = 10\%$; $Z_2 = 100$; $Z_1 = 100$				
0.781	0.781	4.139×10^3	1.00	1.00
0.557	0.557	2.955×10^4	0.10	1.00
0.781	0.781	4.146×10^3	1.00	0.10
0.545	0.557	2.957×10^4	0.10	0.10
0.506	0.506	2.710×10^5	0.01	1.00
0.207	0.506	2.783×10^5	0.01	0.01
$I_0 = 0.1\%$; $Z_2 = 10^3$; $Z_1 = 1.0$				
0.048	0.506	2.771×10^5	0.01	1.00
0.018	0.506	2.752×10^5	0.01	0.10
0.015	0.506	2.731×10^5	0.01	0.01
$I_0 = 0.1\%$; $Z_2 = 10^3$; $Z_1 = 100$				
0.400	0.506	2.809×10^5	0.01	1.00
0.117	0.506	2.855×10^5	0.01	0.10
0.032	0.506	2.749×10^5	0.01	0.01

[a] R. A. McKee and J. P. Stark, ibid.

to zero independent of jump frequencies. This is evidenced by the fact that the partial correlation factors for the two configurations for the P_0 complex go to $\pm \frac{1}{3}$ for large binding energies, as calculated by an independent and exact calculation. As a consequence, the total correlation factor approaches zero whenever $e^{-E_B/K_B T} = e^{-E_V/K_B T} \to 0$ and $Z2 \gg Z1$.

The next observation from the table is that the diffusivity is relatively insensitive to solute concentration. The calculations from this model based on single solute interactions do not indicate any appreciable solute concentration dependence. The main factor influencing solute diffusivity at infinite dilution is the binding energy between the solute and vacancy.

Table 5.8 Solvent self-diffusion

f_T	$I_0 \doteq 0.1\%$ $\langle W \rangle f_T$	X	Y
$Z_2 = 1.0;\ Z_1 = 1.0$			
0.781	4.139×10^3	1.0	1.0000
0.781	4.137×10^3	1.0	0.0100
0.781	4.137×10^3	1.0	0.0001
$Z_2 = 10^3;\ Z_1 = 10^3$			
0.781	4.139×10^3	1.0	1.0000
0.779	4.139×10^3	1.0	0.0100
0.731	4.142×10^3	1.0	0.0001
$Z_2 = 10^6;\ Z_1 = 10^6$			
0.781	4.139×10^3	1.0	1.0000
0.779	4.152×10^3	1.0	0.0100
0.690	4.922×10^3	1.0	0.0001

Changing the factor $e^{-E_B/K_B T}$ from 1 to 0.1 gives one order of magnitude increase in the diffusivity.

This is an indication that for sufficiently dilute alloys, the solute diffusivity is strongly dependent on the probability of a vacancy being within the nearest neighbor shell of the solute atoms. In more concentrated alloys solute clusters must be considered, as in the previous section. Therein an increased vacancy fraction did not appreciably influence solute diffusivity; however, solute concentration was a factor.

Another important factor illustrated by these numerical results is the influence of divacancies. By increasing the positive interaction between vacancies, as demonstrated in the parameter Y, the average jump frequency is seen to increase with Y changes, whereas the diffusivity is unaffected. The explanation for this is the corresponding decrease in the total correlation factor for these cases. The numerical results can reflect solvent self-diffusion if the values for the parameter x are restricted to 1, and only values of $Z_2 = Z_1$ are considered. Table 5.8, is a lists these cases. In the work by Rothman, Peterson, and Robinson[8], a temperature dependence of the correlation factor is reported by isotope experiments with pure silver. A deviation from the value of $f = 0.781$ is shown to exist. This effect is evidenced in this calculation, as presented in Table 5.8. The effect of an increased jump rate into a divacancy is the increase in the average jump frequency, leaving the diffusivity unchanged by decreasing the total correlation factor. These results are not inconsistent with the results reported by Rothman et al. for self-diffusion in silver.

In some experimental work curvature in the Arrhenius plot of diffusivity has been believed to exist. It has been speculated that such results are due to the influence of divacancies. These calculations strongly suggest that although the above reasoning is valid for self-diffusion, it is not the case for a substitutional solute of a different element in solution. The latter effect is due to a temperature dependence of the correlation factor for single solute-single vacancy interactions and does not show the presence of divacancies. Choosing values of $Z_2 = Z_1 = Y = 1$ and varying X from 1 to 0.01 in Table 5.7, the effective jump frequency, $\langle W \rangle f$, changes from 4.14×10^3 to 2.67×10^5. In these calculations, it was assumed that $w_1 = w_2 = w_4 = w_3 e^{E_B/K_B T}$. With these assumptions, the correlation factor for the C complex is defined as $F_c = (2 + 5.15X)/(4 + 5.15X)$, where $X = e^{-E_B/K_B T}$. If the values of $X = 1$ and 0.01 are used to evaluate f_c, $f_{c_{x=1}} = 0.781$ and $f_{c_{x=0.01}} = 0.506$. On this basis the changes in $\langle W \rangle f$ are due to changes in the C complex concentration. If X were varied over this range and E_B held constant, the numbers for $\langle W \rangle f$ would effectively be the temperature variation of the diffusivity. The $\ln [\langle W \rangle f]$ would clearly not be linear in $1/T$. The values of the average jump frequency, $\langle W \rangle$, are expected to be linear in $1/T$; therefore, this nonlinearity is interpreted as a correlation factor variation with temperature. If values of Z_2, Z_1, and Y, which represent divacancy effects are taken, the overall value of $\langle W \rangle f$ changes only slightly. For example, when $Z_2 = Z_1 = 100$ and $Y = 0.01$, varying the value of X between 1 and 0.01 changes $\langle W \rangle f$ from 4.18×10^3 to 2.72×10^5. These values are comparable to the previously cited values of 4.14×10^3 and 2.67×10^5, where divacancies are not an influence.

From these tables, it is clear that ordinary experimental determination of the solute diffusion coefficient in dilute alloys cannot show an effect of divacancies. To investigate divacancy influence, the correlation factor would be a more adequate determination.

RANDOM WALK CALCULATION OF $\langle W \rangle$, C_α, AND D^9

From the previous two examples, it is apparent that the diffusivity calculation in complicated theories presents two difficulties: the calculation of $\langle W \rangle$ and C_α, as well as the calculation of the partial correlation factors, f_α. The C_α requires the manipulation of a large number of kinetic equations, and the correlation factor requires the manipulation of the matrix A. It would prove very convenient if the C_α could be determined by the matrix A or ones similar to it, particularly if one wishes to determine the tracer mobility. Such a calculation technique is possible using vacancy random walk to solve the kinetic equations. The technique is similar to that used in Chapter 3 to calculate the diffusivity by random walk. This discussion is a matrix generalization of that presentation.

To follow the fluctuation leading to tracer diffusion and subsequent vacancy configuration equilibrium at the tracer, one views the tracer at an arbitrary site in a cubic crystal with the X direction oriented perpendicular to (100) crystal planes. With small solute concentrations, the tracer may be assumed to be at some distance from defect sources and sinks. One simple way to view the situation is for the tracer again to be the center of a sphere whose surface contains a uniform distribution of sources and sinks. With a tracer jump distance of λ, the x projection of the tracer jump is b. The defect starts its motion at the source and migrates up to the tracer. One averages all subsequent returns of this particular defect to the tracer, and it follows that one must specify that the defect has not previously arrived at a site where it can exchange with the tracer. Defect equilibrium will be assumed to exist on the surface of the sphere of sources.

In general, there is a number of tracer-defect configurations that may lead to a single tracer jump with nonzero x projection. These are again denoted as α, $\alpha = 1, \cdots N$. There are two configurations for a bound divacancy and the mirror image of these two; hence $N = 2$ for jumps in the positive x direction. When the divacancy can dissociate there are 12 configurations. Of these configurations, nine have distinguishable mirror images and the rest do not. One must distinguish between jumps in the positive and negative x directions; the 12 two-vacancy solute configurations can lead to positive jumps. The nine distinguishable image and three indistinguishable image tracer jumps can occur. The defect could cause a tracer jump of (1) $+b$ with an effective probability $P_{\alpha\beta}^a$, or the tracer could move a distance (2) $-b$ with a probability of $P_{\alpha\beta}^{11}$.

If we assume that (1) occurs, the next tracer displacement could give a net displacement of (3) $+2b$ with a probability $P_{\beta\gamma}^{11}$, or a net displacement of (4) $0b$ with a probability of $P_{\beta\gamma}^a$. Otherwise, when (2) occurs the tracer could make a net displacement of (5) $-2b$ with a probability $P_{\beta\gamma}^{11}$ or (6) $0b$ with a probability $P_{\beta\gamma}^a$.

The probability that (1) occurs and is followed by the defect migrating to a sink is evidently

$$ P_{\alpha\beta}^a \left(1 - \sum_\gamma P_{\beta\gamma}^{11} - \sum_\gamma P_{\beta\gamma}^a \right) \qquad 5.41 $$

The term in parenthesis is the probability that no further jumps occur, which implies that the β jump was the only tracer jump.

With N jump configurations, one arranges the elements $P_{\alpha\beta}^a$ into a matrix p^a, $N \times N$. Let 1 be the unit column vector; the relation in Eq. 5.41 may be written as

$$P^a(I - P^{11} - P^a)1 \qquad 5.42$$

where I is the N-dimensional unit matrix. On the other hand, process (2) could occur and be followed by the defect migrating to a sink. The effective probability is

$$P^{11}(I - P^{11} - P^a)1 \qquad 5.43$$

It follows that the components of the vector

$$1^T \nu_{(-)}(P^{11} + P^a)(I - P^{11} - P^a)$$

are the total number of jumps per unit time of type α for a one-jump sequence. The $\nu_{(-)}$ is a diagonal matrix with elements $\nu_{\pi-\alpha}$. Also, the frequency of occurrence of sequences that start as described above, and have n jumps, is Γ_n where

$$\Gamma_n = 1^T \nu_{(-)}(P^{11} + P^a)^n(I - P^{11} - P^a)1 \qquad 5.44$$

The total frequency Γ_T of jumps in these sequences is

$$\Gamma_T = \sum_{n=1}^{\infty} n\Gamma_n = 1^T \nu_{(-)}(P^{11} + P^a)(I - P^{11} - P^a)^{-1}1 \qquad 5.45$$

Then Γ_T can also be expressed as

$$\Gamma_T = \sum_{\alpha=1}^{N} \Gamma_\alpha \qquad 5.46$$

where Γ_α is the frequency of α type jumps from these sequences. When written in a matrix notation similar to Eq. 5.45, Γ_T becomes

$$\Gamma_T = \Gamma \cdot 1 \qquad 5.47$$

where Γ is a row vector having components Γ_α. A matrix expression for Γ and the Γ_α can be found by comparing Eqs. 5.45 and 5.46.

With each jump providing an x-displacement of either $+b$ or $-b$, the diffusion coefficient D can be written as

$$D = \tfrac{1}{2}b^2 \sum_{\alpha=1}^{N} \nu_\alpha f_\alpha$$

where ν_α is the jump frequency for $+b$ jumps of type α, and f_α is the partial correlation factor for jumps of type α. Since only half of the vacancies jumps are included in the sequence appearing in Eq. 5.45. Thus with $+\alpha$ and $-\alpha$ jumps being equally likely to occur,

$$\nu_\alpha = 2\Gamma_\alpha \qquad 5.48$$

Also, from the equations of Howard,[10] one finds that

$$f_\alpha = 1 + 2 \sum_{i=1}^{\infty} \langle x_\alpha x_{\alpha,i} \rangle b^{-2} \qquad 5.2$$

where $x_{\alpha,i}$ is the x-displacement of jump i after a jump of type α, and x_α is the x-displacement of jump α itself. From Eqs. 5.10 and 5.11

$$f_* = 1 + 2(P^{11} - P^a)(I - P^{11} + P^a)^{-1} 1 \qquad 5.49$$

where column vector f_* has components f_α.

If one substitutes $P^{11} + P^a = P_1$ and $P^{11} - P^a = P_2$, then Eqs. 5.45–5.50 yield

$$D = b^2 1^T \nu_{(-)} P_1 (I - P_1)^{-1} (I + P_2)(I - P_2)^{-1} 1 \qquad 5.50$$

To evaluate Eq. 5.50 and thereby the diffusivity, one calculates P^{11} and P^a from Eq. 5.28. The matrix elements $\nu_{\pi-\alpha}$ as contained in $\nu_{(-)}$ are calculated using Eq. 4.61.

When the correlation factor is set equal to unity, the diffusivity divided by b^2 is $\langle W \rangle / 3$. One can set the correlation factor equal to unity in Eq. 5.50 by letting $P_2 = 0$. Then

$$\langle W \rangle = 3 1^T \nu_{(-)} P_1 (I - P_1)^{-1} 1 \qquad 5.51$$

Furthermore, since the diffusivity is given by

$$D = \frac{b^2}{3} \langle W \rangle f \qquad 5.52$$

One can calculate the correlation factor

$$f = \frac{1^T \nu_{(-)} P_1 (I - P_1)^{-1} (I + P_2)(I - P_2)^{-1} 1}{1^T \nu_{(-)} P_1 (I - P_1)^{-1} 1} \qquad 5.53$$

Furthermore, it is apparent that the fraction of jumps of type α, C_α, are given as elements in the row vector $\underline{C}$ (see Eq. 5.12) as

$$\underline{C} = \frac{1^T \nu_{(-)} P_1 (I - P_1)^{-1}}{1^T \nu_{(-)} P_1 (I - P_1)^{-1} 1} \qquad 5.54$$

The main advantage of Eqs. 5.50–5.54 is that one must tabulate the correlation factor because of the size of the matrix A. With the matrix A, one can easily determine the matrix A_3 needed to find $\nu_{\pi-\alpha}$. Thus in addition to the vectors $Q(\beta)$ and $R(\alpha)$, which are needed for the correlation factor, only the vectors T and H, as presented in Eq. 4.61, are required to determine $\langle W \rangle$, f, $\underline{C}$, and D. Furthermore, when one is considering the calculation of the tracer mobility, these equations are needed. In such an instance, the elements of the matrix

$$[P_1(I - P_1)^{-1}]_{\alpha\beta} = Q_0(\beta)(I - A_{20})^{-1}[R(\alpha^I) + R(\alpha)] \qquad 5.55$$

provide the easiest calculation of the terms in Eqs. 5.50–5.54 since one must have the A_2 matrix to calculate the mobility. Furthermore, since the matrix $S_0{}^+$ must be calculated to determine the mobility, one may use Eq. 4.49 to get the matrix f. That is,

$$f = (I + P_2)(I - P_2)^{-1} = I - 2S_0{}^+ \qquad 5.56$$

Then the matrix A will be unnecessary since all the terms in Eqs. 5.50–5.54 can be determined using A_2 and A_3. An example on this computational procedure is presented in the next section.

THERMAL DIFFUSION OF SOLUTE PAIRS IN AN fcc LATTICE[11]

The concentration dependence of the solute diffusivity was investigated in the third section of this chapter. Therein it was shown that the binding between vacancies and solute pairs influences the diffusivity. It is not unreasonable to presume that such effects would also influence the mobility of a solute tracer in either a temperature or electric potential gradient. We investigate such a possibility in this section.

The flux of substitutional solute tracer atoms in an fcc lattice of a dilute binary alloy follows the generalized form of Fick's first law.

$$J = -D\frac{\partial c}{\partial x} + cv_F \qquad 5.57$$

where c is concentration, and v_F is velocity originating from an applied force.

Since diffusion is isotropic in this lattice, one may orient the crystal so that the $+x$ direction is parallel to the applied field and perpendicular to (100) crystallographic planes. The projection on the x axis of any tracer jump via a vacancy mechanism will have a magnitude of $\pm b$ or 0. Only tracer jumps with nonzero x projection contribute to the flux. On the average, the tracer will have an effective jump frequency parallel to the field of ν_{+e} and antiparallel with the field ν_{-e}. The velocity v_F is found, from Chapter 4, as

$$\begin{aligned} v_F &= b(\nu_{+e} - \nu_{-e}) \\ &= 2b(\nu_{+e} - \nu_{+0}) \end{aligned} \qquad 4.51$$

where the subscript 0 is intended to imply the absence of an applied field herein. The second form for Eq. 4.51 is a direct result of the assumption that the velocity is linearly proportional to the applied field.

The mobility of the tracer is found directly from Eq. 4.51. The effective heat of transport for thermal diffusion is defined as

$$Q_{eff} = -\frac{KT^2 v_F}{D \nabla T} \qquad\qquad 5.58$$

Consequently, knowledge of the frequency v_{+e}, with and without the field present, and of the diffusivity, D, permits the evaluation of the effective heat of transport. One could similarly evaluate the effective ionic charge for electromigration, Q, from a similar expression

$$Q = \frac{KT v_F}{DE} \qquad\qquad 5.59$$

given the appropriate information. In both cases, one must calculate the ratio $KT v_F/D$.

For the present problem there are 12 distinguishable configurations of the first-neighbor tracer, vacancy, and second solute atom, and one configuration for the isolated tracer-vacancy pair, which may lead to a tracer displacement of $+b$. Hence in terms of the correlation factor, there are 13 jump configurations. These are given in Table 5.9; it is to be assumed that the tracer and vacancy are isolated from the second solute atom whenever one of the three are not first-nearest neighbor of the other two.

To evaluate Eq. 5.52 and hence the effective heat of transport for thermal diffusion one must originate the set of two vacancy walk matrices, A_2 and A_3. To this end, consider a set of M defect configurations relative to the tracer. In this problem, as given in Table 5.9, $M = 51$ as follows: There are 21 sets of configurations where the vacancy, tracer, or solute are a nearest neighbor to the other two, and the solute and vacancy are not on the same plane as the tracer. There are 21 mirror images of these configurations, and 6 configurations where the solute and vacancy are on the same x plane as the tracer. This totals 48 configurations where the solute is assumed to influence the vacancy motion around the tracer. In addition, once the solute is a second neighbor to the tracer-vacancy pair and does not neighbor the vacancy, the solute is assumed to be random. There are then three additional configuration sets for the isolated tracer-vacancy pair, which have the vacancy to the right, left, and on the same x plane as the tracer and still a neighbor to it. Thus a total of 51 sets of configurations are investigated (see Table 5.9).

Table 5.9 Configuration classes[a]

Configuration (i)	Tracer (0, 0, 0)		Number of Configurations per class	Jump Configurations (α)
	Solute	Vacancy		
1	$(\bar{1}, 1, 0)$	$(1, 1, 0)$	4	1
2	$(\bar{1}, \bar{1}, 0)$	$(1, 1, 0)$	4	2
3	$(\bar{1}, 0, 1)$	$(1, 1, 0)$	8	3
4	$(2, 0, 0)$	$(1, \bar{1}, 0)$	4	4
5	$(2, 2, 0)$	$(1, 1, 0)$	4	5
6	$(2, 1, 1)$	$(1, 1, 0)$	8	6
7	$(1, \bar{1}, 0)$	$(2, 0, 0)$	4	—
8	$(1, 1, 0)$	$(2, 2, 0)$	4	—
9	$(1, 0, 1)$	$(2, 1, 1)$	8	—
10	$(0, \bar{1}, 1)$	$(1, 0, 1)$	8	7
11	$(0, 2, 0)$	$(1, 1, 0)$	4	8
12	$(0, \bar{1}, 1)$	$(1, 1, 0)$	8	9
13	$(0, 1, 1)$	$(1, 2, 1)$	8	—
14	$(1, 0, 1)$	$(1, 1, 0)$	8	10
15	$(1, \bar{1}, 0)$	$(1, 1, 0)$	4	11
16	$(1, 2, 1)$	$(1, 1, 0)$	8	12
17	$(1, 1, 0)$	$(1, 2, 1)$	8	—
18	$(1, 1, 0)$	$(0, 1, 1)$	8	—
19	$(1, \bar{1}, 0)$	$(0, 1, 1)$	8	—
20	$(1, 1, 0)$	$(0, 2, 0)$	4	
21	$(1, 2, 1)$	$(0, 1, 1)$	8	—
$21+I$ is the mirror image of I, $I \leqslant 21$				
43	$(0, 1, 1)$	$(0, 2, 0)$	8	—
44	$(0, 2, 0)$	$(0, 1, 1)$	8	—
45	$(0, 1, 1)$	$(0, 2, 2)$	4	—
46	$(0, 2, 2)$	$(0, 1, 1)$	4	—
47	$(0, \bar{1}, 1)$	$(0, 1, 1)$	8	—
48	$(0, \bar{1}, \bar{1})$	$(0, 1, 1)$	4	—
49	—	$(1, 1, 0)$	4	13
50	—	$(0, 1, 1)$	4	—
51	—	$(\bar{1}, 1, 0)$	4	—

[a] J. P. Stark, *Acta Met.* **22,** 1349 (1974).

Table 5.10 gives some selected matrix elements for the matrices used in this problem. Therein a model is used that parallels the one given in the section on the influence of solute pairs on the diffusivity. In that model $X = e^{-E_B/KT}$ is the solute-vacancy bonding term, $y = e^{-Ev/KT}$ is the solute-solute bonding term, and z is the increased vacancy-tracer jump rate when a second solute is within the first neighbor shell. As an example of this model, the correlation factor for an isolated solute-vacancy pair is $f = (2+7X)/(4+7X)$. The heat of transport results of Howard and Manning, as deduced in Eq. 4.84, for the isolated solute-vacancy pair may be written†

$$Q_{\text{eff}} = q_2 + \frac{3Xq_3 - q_w(4-3X)}{(2+7X)} \qquad 5.60$$

where $q_4 = q_1 = q_w$ by assumption. Compare this with Eq. 4.84. One may use Eq. 5.60 as a basis for the work herein. Thereby, one defines α_i as functions of X, y, and z as defined above. Since one expects Q_{eff} to remain linear in the heats of transport, q_i, (given in Eq. 5.60), which is a consequence of the linearity of the Eq. 4.51 in the force, one may write the effective charge as follows:

$$Q_{\text{eff}} = q_2[1 + \alpha_1(X, y, z)I_0]$$
$$+ \{3Xq_3[1 + \alpha_2(X, y, z)I_0]$$
$$- q_w(4-3X)[1 + \alpha_3(X, y, z)I_0]\}/(2+7X) \qquad 5.61$$

In Eq. 5.61 I_0 is the solute concentration expressed as the mole fraction solute. As indicated by Eqs. 5.58 and 5.59, one must also calculate the diffusion coefficient prior to the determination of Q_{eff}. The diffusion coefficient is also expected to be linear in the concentration. Therefore, Eq. 5.50 leads to the expectation that

$$D = D_0[1 + \alpha_0(X, y, z)I_0] \qquad 5.62$$

and Q_{eff} is determined using Eqs. 4.61, 4.66, 4.67 and 5.50, 5.58, and 5.59.

By taking suitable variations on the vacancy source so as to reflect the number of configurations contained in the 21 sets and the isolated solute-vacancy, and also to reflect solute-solute bonding and solute-vacancy bonding at the source, one may calculate Q_{eff} and D. Having

---

† Here $q_i = q_i^* - \Delta H_i$ where ΔH_i is the vacancy formation energy at a site prior to a jump w_i.

Table 5.10[a] Selected matrix elements in A_2, Q, and R

$[A_2(I, J) = A(I, J) + \delta A(I, J)]$

$B_1 = 1 + 2z + 4X + 5X^2$

$B_2 = 2 + 2Xz + 4X + 4X^2$

$B_3 = yz + 6X + 5$

$B_4 = 2Xz + 8X + 2X^2$

$B_5 = 4 + 7X + yz$

$B_6 = 2Xz + 1 + 5X + 4X^2$

$B_7 = yz + 4 + 7X$

$D1 = 4Xq_3 - 3q_w - yzq_2$

$D2 = 3Xq_3 - 2q_w - yzq_2$

$D3 = 3Xq_3 - yzq_2 - 2q_w$

$D4 = D5 = D6 = 0$

$D7 = D1, \quad D8 = D2, \quad D9 = D3$

$D10 = 4X^2q_3 - 2Xq_3 - 2zq_2$

$D11 = 4X^2q_3 - 2Xq_2 - 2q_w$

$D12 = 3Xq_3 - yzq_2 - 2q_w$

$D13 = D12$

$D14 = 3X^2q_3 - zq_2 - 2q_w$

$D15 = 3Xq_3 - 2q_w - yzq_2$

$D16 = 3X^2q_3 - Xq_3 - q_w - Xzq_2$

$D17 = 3Xq_3 - 2q_w - yzq_2$

$D18 = -D14, \quad D19 = -D13, \quad D20 = -D15$

$D21 = -D16$

$A(41, 1) = 1/B_3$	$\delta A(41, 1) = (q_w + D1/B_3)/B_3$
$A(20, 22) = 1/B_3$	$\delta A(20, 22) = -\delta A(41, 1)$
$A(39, 1) = 1/B_3$	$\delta A(39, 1) = \delta A(41, 1)$
$A(18, 22) = 1/B_3$	$\delta A(18, 22) = -\delta A(41, 1)$
$A(3, 1) = 1/B_3$	$\delta A(3, 1) = D1/B_3^2$
$A(24, 22) = 1/B_3$	$\delta A(24, 22) = -\delta A(3, 1)$
$A(40, 2) = 1/B_5$	$\delta A(40, 2) = (q_w + D2/B_5)/B_5$
$A(19, 23) = 1/B_5$	$\delta A(19, 23) = -\delta A(3, 2)$
$A(3, 2) = 1/B_5$	$\delta A(3, 2) = D2/B_5^2$
$A(24, 23) = 1/B_5$	$\delta A(24, 23) = -\delta A(1, 3)$
$A(1, 3) = 2/B_7$	$\delta A(1, 3) = 2D3/B_7^2$
$A(22, 24) = 2/B_7$	$\delta A(22, 24) = -\delta A(1, 3)$
$A(2, 3) = 2/B_7$	$\delta A(2, 3) = \delta A(1, 3)$
$A(23, 24) = 2/B_7$	$\delta A(23, 24) = -\delta A(1, 3)$
$A(39, 3) = 1/B_7$	$\delta A(39, 3) = (q_w + D3/B_7)/B_7$
$A(18, 24) = -1/B_7$	$\delta A(18, 24) = -\delta A(39, 3)$
$A(40, 3) = 1/B_7$	$\delta A(40, 3) = \delta A(39, 3)$

Table 5.10 (*continued*)

$$A(19, 24) = 1/B_7 \qquad \delta A(19, 24) = -\delta A(39, 3)$$
$$A(4, 4) \quad = 2/B_2 \qquad \delta A(4, 4) \quad = 0.0$$
$$\dot A(25, 25) = 2/B_2 \qquad \delta A(25, 25) = 0.0$$
$$A(7, 4) \quad = Xz/B_2 \qquad \delta A(7, 4) \quad = -Xzq_2/B_2$$
$$A(28, 25) = Xz/B_2 \qquad \delta A(28, 25) = -\delta A(7, 4)$$
$$A(8, 5) \quad = Xz/B_4 \qquad \delta A(8, 5) \quad = Xzq_2/B_4$$
$$A(29, 26) = Xz/B_4 \qquad \delta A(29, 26) = -\delta A(8, 5)$$
$$A(9, 6) \quad = Xz/B_6 \qquad \delta A(9, 6) \quad = -Xzq_2/B_6$$
$$A(30, 27) = Xz/B_6 \qquad \delta A(30, 27) = -\delta A(9, 6)$$
$$A(6, 6) \quad = 1/B_6 \qquad \delta A(6, 6) \quad = 0.0$$
$$A(27, 27) = A(6, 6) \qquad \delta A(27, 27) = -\delta A(6, 6)$$

$I, J \leq 21$, $A(I+21, J+21) = A(I, J)$, $\delta A(I+21, J+21) = -\delta A(I, J)$, and so forth.

$Q(\alpha)$

$$Q_0(1, 1) \quad = 4yz/B_3 \qquad \delta Q(1,1) \quad = 4yz(q_2 + D1/B_3)/B_3$$
$$Q_0(2,2) \quad = 4yz/B_5 \qquad \delta Q(2, 2) \quad = 4yz(q_2 + D2/B_5)/B_5$$
$$Q_0(3, 3) \quad = 8yz/B_7 \qquad \delta Q(3, 3) \quad = 8yz(q_2 + D3/B_7)/B_7$$
$$Q_0(4, 4) \quad = 4Xz/B_2 \qquad \delta Q(4, 4) \quad = 4yz(q_2)/B_2$$
$$Q_0(5, 5) \quad = 4Xz/B_4 \qquad \delta Q(5, 5) \quad = 4Xzq_2/B_4$$
$$Q_0(6, 6) \quad = 8Xz/B_6 \qquad \delta Q(6, 6) \quad = 8Xzq_2/B_4$$
$$Q_0(7, 10) \quad = 8z/B_1 \qquad \delta Q(7, 10) \quad = 8z(q_2 + D10/B_1)/B_1$$
$$Q_0(8, 11) \quad = 4Xz/B_2 \qquad \delta Q(8, 11) \quad = 4Xz(q_2 + D11/B_2)/B_2$$
$$Q_0(9, 12) \quad = 8yz/B_7 \qquad \delta Q(9, 12) \quad = 8yz(q_2 + D12/B_7)/B_7$$
$$Q_0(10, 14) = 8z/B_1 \qquad \delta Q(10, 14) = 8z(q_2 + D14/B_1)/B_1$$
$$Q_0(11, 15) = 4yz/B_3 \qquad \delta Q(11, 15) = 4yz(q_2 + D15/B_3)/B_3$$
$$Q_0(12, 16) = 8Xz/B_6 \qquad \delta Q(12, 16) = 8Xz(q_2 + D16/B_6)/B_6$$

$R(\alpha^I)$

$$R(4, 1) \quad = \tfrac{1}{4} \qquad R(5, 2) \quad = \tfrac{1}{4}$$
$$R(6, 3) \quad = \tfrac{1}{8} \qquad R(1, 4) \quad = \tfrac{1}{4}$$
$$R(2, 5) \quad = \tfrac{1}{4} \qquad R(3, 6) \quad = \tfrac{1}{8}$$
$$R(14, 7) \quad = \tfrac{1}{8} \qquad R(15, 8) \quad = \tfrac{1}{4}$$
$$R(16, 9) \quad = \tfrac{1}{8} \qquad R(10, 10) = \tfrac{1}{8}$$
$$R(11, 11) = \tfrac{1}{4} \qquad R(12, 12) = \tfrac{1}{8}$$

---

[a] J. P. Stark, *Acta Met.* **22,** 1349 (1974).

made these calculations, one may determine the α_i's in Eq. 5.61 as follows:

$$\alpha_1 = \frac{1}{I_0}\left[\frac{\partial Q_{eff}}{\partial q_2} - 1\right] \qquad\qquad 5.63$$

$$\alpha_2 = \frac{1}{I_0}\left\{\left[\frac{(2+7X)}{3X}\right]\frac{\partial Q_{eff}}{\partial q_3} - 1\right\} \qquad\qquad 5.64$$

$$\alpha_3 = \frac{1}{I_0}\left\{-\left(\frac{2+7X}{4-3X}\right)\frac{\partial Q_{eff}}{\partial q_w} - 1\right\} \qquad\qquad 5.65$$

Equations 5.63–5.65 are easily performed on the computer once one has made the calculations indicated for Eq. 5.61. In addition, one may calculate the correlation factor, f, as given in Eq. 5.53.

Table 5.11 presents a tabulation of the parameters α_i, $i = 1, 2, 3$, that are dependent on the jump parameters X, y, z. As a check on the α_1, α_2, and α_3 values, this table presents values of Q_{eff} evaluated for a 1% alloy, and f for a 1% alloy.

Table 5.11   Solute pair concentration parameters[a]

z	X	y	$f_x(I_0=0.01)$	Q_{eff} $(I_0=0.01, q_i=1)$	α_1	α_2	α_3
0.01	0.1	0.1	0.57	−0.74	0.08	895	0.00
0.01		1.0	0.57	0.74	0.01	899	0.00
0.01		10.0	0.57	0.74	0.00	899	0.00
0.01	1.0	0.1	0.82	1.22	0.01	−0.03	0.00
0.01		1.0	0.82	1.22	0.00	−0.03	0.00
0.01		10.0	0.82	1.22	0.00	−0.04	0.00
0.01	10.0	0.1	0.97	1.40	0.00	−89.9	0.00
0.01		1.0	0.97	1.40	0.00	−90.0	0.00
0.01		10.0	0.97	1.40	0.00	−90.0	0.00
0.10	0.1	0.1	0.57	0.71	0.21	869	−0.20
0.10		1.0	0.57	0.73	0.00	896	−0.08
0.10		10.0	0.57	0.74	−0.02	898	0.00
0.10	1.0	0.1	0.82	1.22	0.02	−0.05	0.00
0.10		1.0	0.82	1.22	0.00	−0.06	0.00
0.10		10.0	0.82	1.22	−0.00	−0.06	0.00
0.10	10.0	0.1	0.97	1.40	0.00	−89.9	0.00
0.10		1.0	0.97	1.40	0.00	−90.0	0.00
0.10		10.0	0.97	1.40	0.00	−90.0	0.00
1.0	0.1	0.1	0.49	0.67	−0.68	827	−1.30
1.0		1.0	0.56	0.73	−0.26	869	−0.50

Table 5.11 *(continued)*

z	X	y	$f_x(I_0 = 0.01)$	Q_{eff} $(I_0 = 0.01, q_i = 1)$	α_1	α_2	α_3
1.0		10.0	0.57	0.74	−0.18	895	0.00
1.0	1.0	0.1	0.82	1.22	−0.17	−0.34	−5.87
1.0		1.0	0.82	1.22	−0.01	−0.20	−1.37
1.0		10.0	0.82	1.22	−0.00	−0.16	−0.61
1.0	10.0	0.1	0.97	1.40	0.02	−89	−0.41
1.0		1.0	0.97	1.40	0.01	−89	−0.40
1.0		10.0	0.97	1.40	0.01	−89	−0.38
10.0	0.1	0.1	0.20	0.52	−12.4	789	−1.98
10.0		1.0	0.49	0.72	−1.58	876	−1.19
10.0		10.0	0.55	0.74	−0.23	896	−0.33
10.0	1.0	0.1	0.72	1.19	−4.11	−3.05	−20.8
10.0		1.0	0.77	1.22	−0.25	−1.00	−6.01
10.0		10.0	0.78	1.22	−0.09	−0.32	−1.35
10.0	10.0	0.1	0.96	1.44	0.32	−81.5	−1.15
10.0		1.0	0.96	1.42	0.27	−84.6	−1.38
10.0		10.0	0.93	1.41	0.23	−87.4	−1.40
100.0	0.1	0.1	0.03	−0.75	−135	744	−2.32
100.0		1.0	0.19	0.61	−12.1	874	−1.37
100.0		10.0	0.38	0.73	−0.89	896	−0.13
100.0	1.0	0.1	0.28	0.94	−30.7	−6.18	−37.00
100.0		1.0	0.45	1.21	−0.73	−1.75	−9.25
100.0		10.0	0.51	1.22	1.22	0.41	−0.38
100.0	10.0	0.1	0.77	1.33	−25.2	−46.6	−1.10
100.0		1.0	0.66	1.38	−9.76	−68.9	−1.63
100.0		10.0	0.56	1.43	2.15	−85.8	−2.11

[a] J. P. Stark, *Acta Met*, **22**, 1349 (1974).

Table 5.12 Dependence on vacancy binding[a]

X	$f_x(I_0 = 0.01)$	Q_{eff} $(I_0 = 0.01, q_i = 1)$	α_1	α_2	α_3
$y = 0.1$; $z = 3.16$					
0.147	0.44	0.72	−3.67	521	−2.32
0.215	0.52	0.82	−3.69	325	−2.97
0.316	0.60	0.92	−3.42	192	−3.62
0.464	0.67	1.03	−2.83	102	−4.44
0.681	0.74	1.13	−2.01	40.6	−6.13
1.000	0.80	1.22	−1.17	−1.21	−12.60

(continued overleaf)

173

Table 5.12 (*continued*)

X	$f_x(I_0=0.01)$	Q_{eff} $(I_0=0.01, q_i=1)$	α_1	α_2	α_3
1.470	0.85	1.29	−0.51	−30.2	33.10
2.150	0.89	1.34	−0.11	−50.5	5.39
3.160	0.92	1.37	0.06	−65.0	2.07
4.640	0.94	1.39	0.11	−75.1	0.67
6.810	0.96	1.41	0.11	−82.1	−0.13
10.000	0.97	1.41	0.10	−87.0	−0.63
$y=0.1$; $z=10.0$					
0.147	0.27	0.62	−11.5	506	−2.78
0.215	0.35	0.74	−10.3	313	−3.73
0.316	0.44	0.86	−8.94	183	−4.92
0.464	0.53	0.98	−7.37	95.3	−6.60
0.681	0.63	1.09	−5.75	36.4	−9.73
1.000	0.72	1.19	−4.12	−3.05	−20.80
1.470	0.80	1.28	−2.55	−29.5	56.80
2.150	0.85	1.35	−1.22	−47.5	10.30
3.160	0.90	1.39	−0.33	−60.3	4.76
4.640	0.93	1.42	0.15	−69.7	2.07
6.810	0.95	1.43	0.31	−76.6	0.23
10.000	0.96	1.44	0.32	−81.5	−1.15
$y=0.1$; $z=31.6$					
0.147	0.12	0.35	−36.7	488	−3.12
0.215	0.17	0.51	−31.0	301	−4.39
0.316	0.23	0.69	−24.9	174	−6.07
0.464	0.32	0.86	−19.1	89.2	−8.58
0.681	0.42	1.01	−14.3	32.5	−13.3
1.000	0.52	1.13	−10.7	−4.92	−29.0
1.470	0.62	1.23	−7.99	−29.0	78.2
2.150	0.72	1.32	−5.86	−44.2	14.1
3.160	0.80	1.39	−3.99	−53.7	7.23
4.640	0.85	1.44	−2.48	−60.1	4.24
6.810	0.89	1.47	−1.54	−65.0	1.75
10.000	0.91	1.48	−1.06	−68.6	−0.75
$y=0.1$; $z=100$					
0.147	0.05	−0.46	−115	473	−3.48
0.215	0.07	−0.14	−94.5	289	−5.02
0.316	0.09	0.19	−73.5	166	−7.13
0.464	0.14	0.50	−54.5	83.4	−10.4
0.681	0.20	0.75	−40.0	29.0	−16.6
1.000	0.28	0.94	−30.7	−6.18	−37.0
1.470	0.36	1.06	−26.2	−28.0	99.2
2.150	0.47	1.14	−25.2	−46.7	7.54

Table 5.12 (*continued*)

X	$f_x(I_0 = 0.01)$	Q_{eff} $(I_0 = 0.01, q_i = 1)$	α_1	α_2	α_3
3.160	0.57	1.20	−25.2	−48.6	4.50
4.640	0.65	1.25	−25.5	−48.1	2.89
6.810	0.72	1.30	−25.5	−48.1	2.89
10.000	0.77	1.33	−25.2	−46.6	0.91
$y = 10$; $z = 3.16$					
0.147	0.59	0.81	−0.17	579	−0.24
0.215	0.63	0.89	−0.15	362	−0.26
0.316	0.67	0.98	−0.14	215	−0.30
0.464	0.71	1.07	−0.12	115	−0.38
0.681	0.76	1.15	−0.10	46.3	−0.54
1.000	0.80	1.22	−0.08	−0.25	−1.07
1.470	0.85	1.28	−0.05	−31.8	2.29
2.150	0.88	1.32	−0.02	−53.3	1.37
3.160	0.91	1.35	0.00	−67.8	−0.21
4.640	0.93	1.37	0.03	−77.6	−0.45
6.810	0.95	1.39	0.04	−84.2	−0.66
10.000	0.96	1.41	0.05	−88.6	−0.86
$y = 10$; $z = 10$					
0.147	0.57	0.81	−0.21	579	−0.25
0.215	0.61	0.89	−0.19	362	−0.28
0.316	0.65	0.99	−0.17	215	−0.34
0.464	0.69	1.07	−0.15	115	−0.44
0.681	0.73	1.15	−0.12	46.1	−0.65
1.000	0.77	1.22	−0.09	−0.32	−1.35
1.470	0.81	1.27	−0.04	−31.9	3.03
2.150	0.85	1.32	0.02	−53.2	0.23
3.160	0.87	1.35	0.08	−67.6	−0.26
4.840	0.89	1.38	0.14	−77.2	−0.64
6.810	0.91	1.39	0.19	−83.4	−1.04
10.000	0.93	1.41	0.23	−87.4	−1.49
$y = 10$; $z = 31.6$					
0.147	0.52	0.81	−0.34	579	−0.24
0.215	0.55	0.89	−0.29	362	−0.29
0.316	0.59	0.99	−0.23	215	−0.35
0.464	0.62	1.07	−0.16	114	−0.47
0.681	0.66	1.15	−0.08	46.1	−0.71
1.000	0.69	1.22	0.01	−0.36	−1.47
1.470	0.72	1.28	0.11	−31.9	3.33
2.150	0.74	1.32	0.24	−53.2	0.25
3.160	0.76	1.36	0.37	−67.5	−0.29
4.640	0.78	1.38	0.52	−76.9	−0.73

(*continued overleaf*)

Table 5.12 (*continued*)

X	$f_x(I_0 = 0.01)$	Q_{eff} $(I_0 = 0.01, q_i = 1)$	α_1	α_2	α_3
6.810	0.79	1.40	0.64	−82.9	−1.26
10.000	0.80	1.42	0.73	−86.4	−1.93
$y = 10;\ z = 100$					
0.147	0.40	0.81	−0.78	579	−0.24
0.215	0.43	0.89	−0.61	362	−0.29
0.316	0.45	0.98	−0.40	215	−0.36
0.464	0.48	1.07	−0.16	114	−0.48
0.681	0.50	1.16	0.12	46.1	−0.73
1.000	0.51	1.23	0.41	−0.38	−1.52
1.470	0.52	1.28	0.72	−31.9	3.44
2.150	0.53	1.33	1.06	−53.2	2.43
3.160	0.54	1.37	1.41	−67.5	−0.32
4.640	0.54	1.40	1.76	−76.8	−0.79
6.810	0.55	1.41	2.03	−82.6	−1.35
10.000	0.56	1.43	2.15	−85.8	−2.11

[a] J. P. Stark, *Acta Met.* **23,** 1349 (1974).

Table 5.11 also shows that solute pairs influence the effective heat of transport. The major dependence is found in the independent variable X, and the α_1's show large variation on X. Consequently, Table 5.12 was programmed to keep y and z constant and to illustrate the X dependence.

It is apparent from Tables 5.11 and 5.12 that solute-vacancy binding, $X < 1$, is the major factor causing a concentration dependent heat of transport.

REFERENCES

1. R. E. Howard, *Phys. Rev.* **144,** 650 (1966).
2. A. B. Lidiard, *Phil. Mag.* **5,** 1170 (1960).
3. R. E. Howard and J. R. Manning, *Phys. Rev.* **154,** 561 (1967).
4. R. E. Howard and J. R. Manning, ibid.
5. J. P. Stark, *J. Appl. Phys.* **43,** 4404 (1972).
6. R. A. McKee and J. P. Stark, *Phys. Rev.* **B7,** 613 (1973).
7. R. A. McKee and J. P. Stark, ibid.
8. S. J. Rothman, N. K. Peterson, and J. T. Robinson, *Phys. Stat. Solids* **39,** 635 (1970).
9. J. P. Stark, *Acta Met.* **22,** 533 (1974).
10. R. E. Howard, *Phys. Rev.* **144,** 650 (1966).
11. J. P. Stark, *Acta Met.* **22,** 1349 (1974).

CHAPTER 6

MATRIX EQUATIONS FOR UNEQUAL JUMP DISTANCES

THE DIFFUSIVITY

In Chapters 3, 4, and 5, the equations for the tracer diffusion coefficient and velocity in an applied field were developed for the situation in which the x projection of the jump distance was $\pm b$, where b is constant. Clearly, the symmetry necessary for this single jump distance is not always present. In what follows, the diffusion coefficient and velocity in an applied field are developed into a set of matrix equations capable of describing the diffusion flux when the necessary mirror symmetry among the complexes is not present.

In the present development multiple jump distances are allowed and, consequently, we ignore the possibility of reducing the size of the matrix S_{0+} to exclude the mirror symmetry. Every configuration that can move the tracer along the x axis is included in the set α. We start with the diffusivity. Let x_α be the projection of the jump configuration α along the x axis; x_α is positive or negative, there are a number of jump types, α, $1 \leq \alpha \leq r$. Herein, these jump types include all distinguishable tracer defect configurations that lead to a nonzero tracer displacement of magnitude x_α, where x_α is positive for some α and negative for others. Then, if Γ_α is the average jump rate for a tracer jump of type α, one may write the diffusivity

$$D = \tfrac{1}{2} \sum_{\alpha=1}^{r} \Gamma_\alpha x_\alpha^2 f_\alpha \qquad\qquad 6.1$$

177

where f_α is the partial correlation factor for jumps of type α. These partial correlation factors are defined as follows:

$$f_\alpha = 1 + 2 \sum_{m=1}^{\infty} \frac{\langle x_\alpha x_{a,m} \rangle}{x_\alpha^2} \qquad 6.2$$

where the subscript α, m denotes the mth tracer jump following one of type α.

With the above, the diffusion coefficient calculation reduces to the determination of Γ_α and f_α for all α. To facilitate the determination of these quantities, one defines a set of probabilities as follows: $P_{m\alpha\beta}$ is the probability that the mth tracer jump following one of type α will be of type β. With this definition, Eq. 6.2 may be written as

$$f_\alpha = 1 + 2 \sum_{m=1}^{\infty} \sum_{\beta=1}^{r} \frac{P_{m\alpha\beta} x_\beta}{x_\alpha} \qquad 6.3$$

Now one may develop a recurrence formula for the $P_{m\alpha\beta}$ in terms of the $P_{*\alpha\beta}$, where the asterisk denotes the first jump, $m = 1$, as follows: Evidently,

$$P_{m+1\alpha\beta} = \sum_{\gamma=1}^{r} P_{m\alpha\gamma} P_{*\gamma\beta}. \qquad 6.4$$

Consequently, if one writes $P_{m\alpha\beta}$ as an $r \times r$ matrix, P_m.

$$P_m = (P_*)^m \qquad 6.5$$

Letting B be a diagonal matrix with elements x_α, and 1 be the r component unit column vector, one finds, on combining Eqs. 6.3, 6.4, and 6.5 that the column vector

$$\begin{pmatrix} x_1^2 f_1 \\ x_2^2 f_2 \\ \cdot \\ \cdot \\ \cdot \\ \cdot \\ x_r^2 f_r \end{pmatrix} = B\left(I + 2 \sum_{m=1}^{\infty} P_m \right) B 1$$

$$= B[I + 2P_*(I - P_*)^{-1}] B 1 \qquad 6.6$$

when I is the $r \times r$ unit matrix. Then with Γ as a row vector with elements Γ_α, the diffusion coefficient is found from Eqs. 6.1 and 6.6 as

$$D = \tfrac{1}{2} \Gamma B [I + 2P_*(I - P_*)^{-1}] B 1$$

$$= \tfrac{1}{2} \Gamma B (I + P_*)(I - P_*)^{-1} B 1 \qquad 6.7$$

The calculation of the average jump rate vector, Γ, may be performed with the assumption that the defect complexes causing tracer migration are dilute. This evolves into the assumption that a defect complex, such as divacancy, arrives at the tracer and causes a sequence of correlated jumps prior to the arrival of another complex. Under this circumstance the average jump rate will reflect the average rate of arrival of defect complexes at the tracer times the effective probability that tracer jumps can occur prior to defect randomization. Since there are a number of jump configurations α, and since the probability $P_{m\alpha\beta}$ is defined for the β jump following an α jump, it is convenient to define the defect arrival rate ν_α as the effective defect arrival rate at the configuration immediately following an α-type jump. It is assumed that the defect arrives at this configuration without having arrived at any other jump configuration. As before, the configuration following an α-type jump is not an α configuration. Previously the configuration could possibly be a mirror image of α, and thereby be contained within the class of configurations designated by α. Here, however, positive and negative jump configurations are given a different jump configuration designation, so that the ν_α is the formation rate of γ configuration that follows an α-type jump.

Following the arrival of the defect at the configuration following an α jump, the defect could cause one β type jump with the actual jump probability $P'_{\alpha\beta}$. If the defect then randomizes or migrates to a sink, the process is over. Thus the rate at which one β jump occurs is

$$\nu_\alpha P'_{\alpha\beta}\left(1 - \sum_\gamma P_{*\beta\gamma}\right) \qquad 6.8$$

where the last term reflects the defect making no further jumps so that it has gone to its sink. Then P' is not equal to P_*.

If one writes ν_α as a diagonal $r \times r$ matrix, ν, with elements ν_α, and if 1^T is the r component unit row vector, then the elements in Eq. 6.8 are found in the row vector

$$1^T \nu P'(I - P_*) \qquad 6.9$$

Thus if one defines Γ_m as a row vector whose elements reflect the frequency of occurrence of an m-jump sequence, then

$$\Gamma_1 = 1^T \nu P'(I - P_*) \qquad 6.10$$

It follows that for m jumps

$$\Gamma_m = 1^T \nu P' P_{m-1}(I - P_*)$$
$$= 1^T \nu P' P_*^{m-1}(I - P_*) \qquad 6.11$$

The total frequency of jumps of all types is given by $\Gamma 1$ as a consequence of Eqs. 6.10 and 6.11. Thus

$$\Gamma 1 = \sum_{m=1}^{\infty} m\Gamma_m = 1^T \nu P'(I - P_*)^{-1} 1 \qquad 6.12$$

However, by definition

$$\Gamma 1 = \sum_{\alpha=1}^{r} \Gamma_\alpha \qquad 6.13$$

where Γ_α is the frequency of jumps from an α-type sequence. Comparing Eqs. 6.12 and 6.13, the row vector Γ is found from

$$\Gamma = 1^T \nu P'(I - P_*)^{-1} \qquad 6.14$$

One may then substitute Eq. 6.14 into Eq. 6.7 to determine the diffusivity as

$$D = \tfrac{1}{2} 1^T \nu P'(I - P_*)^{-1} B(I + P_*)(I - P_*)^{-1} B 1 \qquad 6.15$$

where $P' = [QR]^T$ and $P_* = [Q(I - A)^{-1}R]^T$; Q, A, and R have similar definitions to those given before. The Q is the matrix of tracer jump probabilities, A is the random walk matrix, and R is the matrix of defect occupation probability following an α jump. An alternate formalism is given in Eq. 6.36 for Q, A, and R in which the transpose is not necessary.

To compare Eq. 6.15 to the previous results, one must introduce a mirror symmetry plane for each of the jump configurations. One assumes that the plane containing the tracer is a mirror plane for each jump configuration; each tracer jump configuration α has a mirror image configuration α^I and r is even. Let r be equal to $2N$ and order the jump configurations α so that for $1 \le \alpha \le N$ one has positive jumps, and let $\alpha + N$ be the reverse image jump in the negative X direction. Thus $X_\alpha = -X_{\alpha+N}$ $P_{m\alpha\beta} = P_{m,\alpha+N,\beta+N}$ and $P_{m\alpha+N,\beta} = P_{m\alpha\beta+N}$ are obvious consequences of this configurational ordering.

To facilitate the comparison define an $N \times N$ matrix, P^{11}, with elements $P^{11}_{\alpha\beta}$ as the effective probability of a tracer jump of type β that is parallel to the previous jump of type α. Also, define an $N \times N$ matrix, P^a, with elements $P^a_{\alpha\beta}$, which are the effective probabilities of a tracer jump of type β that is antiparallel to the previous jump of type α. Finally, subdivide ν_α into two groups of N elements each; those elements are $\nu_{-\alpha}$ where the defect arrives at a configuration following a jump in the negative X direction from a configuration α^I. The defect is thus in a position for a positive jump. Also, $\nu_{+\alpha}$ include the defects that are in the position for a negative jump since the position is one following a positive jump from configuration α.

We distinguish these arrival rates of the defect at the tracer from those rates, $\nu_{\pi \pm \alpha}$, described in the previous chapters. There is a topological problem in the previous definition that excludes the calculation of diffusion by divacancies that may dissociate. To understand the difference, define $\nu_{\pi i}$ as the arrival rate of the defect at complex i, which is not a tracer jump configuration. The $\psi_{i\alpha}$ is the probability of arriving at the tracer jump configuration α without arriving at any other jump configuration. We assume that the entire set of tracer jump configurations is included in this latter probability. This is not true for the divacancy that can dissociate. Equation 6.15 includes the correct arrival rate for the divacancy that can dissociate. To understand this, however, one must follow the random defect walk, as illustrated at the end of this chapter. One also defines $U_{\alpha\beta}$ as the effective probability that the defect can make a transition from an α tracer jump configuration to a β tracer jump configuration. Then, roughly speaking one has

$$\sum \nu_\alpha P' = \sum \nu_{\pi i} \psi_{i\alpha} U_{\alpha\beta} P'_{\beta\gamma} = \sum (\nu_{\pi-\beta} + \nu_{\pi+\beta}) P_{*\beta\gamma}$$

and

$$\sum_\beta U_{\alpha\beta} P'_{\beta\gamma} = P_{*\alpha\gamma}$$

We avoid the above divacancy problem and use the latter nomenclature. To accomplish this, write $\nu_{\pi-\alpha}$ into an $N \times N$ diagonal matrix, $\nu_{(-)}$, and $\nu_{\pi+\alpha}$ into a similar matrix, $\nu_{(+)}$.

Thus for example

$$\nu_\pi = \nu_{11} + \nu_{22} \qquad\qquad 6.16$$

where

$$\nu_{11} = \begin{pmatrix} \nu_{(-)} & 0 \\ 0 & 0 \end{pmatrix}$$

$$\nu_{22} = \begin{pmatrix} 0 & 0 \\ 0 & \nu_{(+)} \end{pmatrix} \qquad\qquad 6.17$$

Since the elements are ordered as assumed, it is obvious that

$$P^{11}_{\alpha\beta} = P_{*\alpha\beta} = P_{*\alpha+N,\beta+N} \qquad\qquad 6.18$$

for $\alpha, \beta < N$. Also,

$$P^a_{\alpha\beta} = P_{*\alpha+N\beta} = P_{*\alpha\beta+N} \qquad\qquad 6.19$$

for $\alpha, \beta < N$. From Eqs. 6.18 and 6.19, one can show by expansion that

$$BP^m B1 = B \begin{pmatrix} P^{11} - P^a & 0 \\ 0 & P^{11} - P^a \end{pmatrix}^m B1 \qquad\qquad 6.20$$

Let d_0 be a diagonal matrix with elements X_α and define

$$V = V_{11} + V_{22}$$

where

$$V_{11} = \begin{pmatrix} P^{11} - P^a & 0 \\ 0 & 0 \end{pmatrix}$$

$$V_{22} = \begin{pmatrix} 0 & 0 \\ 0 & P^{11} - P^a \end{pmatrix} \qquad 6.21$$

With this nomenclature, Eq. 6.20 implies that the partial correlation factors times the appropriate jump distances can be written as,

$$B(I + P_*)(I - P_*)^{-1} B1 = d_0(I + V)(I - V)^{-i} d_0 1 \qquad 6.22$$

Furthermore, if w is defined as $w = w_{11} + w_{22}$ where

$$w_{11} = \begin{pmatrix} P^{11} + P^a & 0 \\ 0 & 0 \end{pmatrix}$$

and

$$w_{22} = \begin{pmatrix} 0 & 0 \\ 0 & P^{11} + P^a \end{pmatrix} \qquad 6.23$$

then the jump rates Γ can be shown by similar expansion to be

$$\Gamma = 1^T \nu P_*(I - P_*)^{-1} = 1^T [\nu_{(-)} + \nu_{(+)}] w(I - w)^{-1} \qquad 6.24$$

If one combines Eqs. 6.22 and 6.23, the diffusion coefficient is found to be

$$D = \tfrac{1}{2} 1^T \nu_{11} w_{11}(I - w_{11})^{-1} d_{11}(I + V_{11})(I - V_{11})^{-1} d_{11} 1$$
$$+ \tfrac{1}{2} 1^T \nu_{22} w_{22}(I - w_{22})^{-1} d_{22}(I + V_{22})(I - V_{22})^{-1} d_{22} 1$$

and

$$D = 1^T \nu_{11} w_{11}(I - w_{11})^{-1} d_{11}(I + V_{11})(I - V_{11})^{-1} d_{11} 1 \qquad 6.25$$

by symmetry. In Eq. 6.25

$$d_0 = d_{11} + d_{22}$$

where

$$d_{11} = \begin{pmatrix} d' & 0 \\ 0 & 0 \end{pmatrix}$$

$$d_{22} = \begin{pmatrix} 0 & 0 \\ 0 & d' \end{pmatrix}$$

and d' is the first N elements in d.

If all the elements in d' are equal to b, one arrives at the result given in Eq. 5.50. Hence

$$D = b^2 1^T \nu_{(-)}(P^{11} + P^a)(I - P^{11} - P^a)^{-1}(I + P^{11} - P^a)(I - P^{11} + P^a)^{-1}1$$

$$5.50$$

Equation 6.25 can also be used to evaluate the correlation factor for comparison with Howard's[†] result. In Eq. 6.25 the diffusion coefficient would become uncorrelated in the situation that $V_{11} = 0$, so that the correlation factor f is given by

$$f = \frac{D}{1^T \nu_{11} w_{11}(I - w_{11})^{-1} d_{11}^2 1} \qquad 6.26$$

If C_α is the fraction of jumps of type α, it is obvious that the row vector C with elements C_α are equal to $\Gamma_\alpha / \sum_\alpha \Gamma_\alpha$. Hence

$$C = \frac{1^T \nu_{11} w_{11}(I - w_{11})^{-1}}{1^T \nu_{11} w_{11}(I - w_{11})^{-1}1} \qquad 6.27$$

In Howard's nomenclature

$$P^{11} - P^a = T_1$$

$$b = \frac{Cd'}{[C(d')^2]1}$$

and

$$\underline{d} = d'1$$

Using this nomenclature in combination with Eqs. 6.25, 6.26, and 6.27 yields

$$f = 1 + 2bT_1(I - T_1)^{-1}\underline{d} \qquad 6.28$$

which is Howard's result for the correlation factor.

TRACER DRIFT VELOCITY

Since the tracer drift velocity is a vector, its motion along the three perpendicular directions in space completely specify the motion. Consequently, one may find a crystal orientation in which the tracer motion in these three directions can be described by a minimal number of tracer-defect configurations that lead to tracer jumps. As before, one must consider both tracer jumps in the positive and negative direction along

[†] R. E. Howard, *Phys. Rev.* **144**, 650 (1966).

these principal orientations, and hence one may assume the field is applied in the positive x direction. All tracer-defect jump configurations for both the positive and negative jump form separate jump configurations. One may order these as α, $\alpha = 1, \cdots K, \cdots r$, where there are K positive and $r - K$ negative jump configurations. Each jump configuration will lead to an x component of the tracer displacement of magnitude x_α.

As an example, consider an fcc crystal with unit cell dimensions of $2b$, oriented with the x direction perpendicular to (100) planes, with a tracer moving by a bound divacancy. There are two configurations of the tracer and divacancy that can lead to a jump of $x_\alpha = +b$, and two of $x = -b$. If on the other hand one had chosen the x direction to be perpendicular to (110) planes, there would be a jump configuration in the plus x direction of $b\sqrt{2}$, and one with $b\sqrt{2}/2$. There would also be the negative of these jumps. Other crystal orientations would lead to a larger value for r, and, as will be seen below, this will needlessly increase the matrix dimensionality.

One must distinguish the possibility of having a positive jump followed by a positive jump relative to the field. Let a $K \times K$ submatrix with elements $P_{\alpha\beta}^{++}$ denote the probability that a jump of type α in the positive direction is followed by a jump of type β in the positive direction. Define a submatrix of dimensions $K \times (r - K)$ with elements $P_{\alpha\beta}^{+-}$, which are the probabilities that a plus jump of type α is followed by a minus jump of type β. There will also be submatrices of dimension $(r - K) \times K$ with elements $P_{\alpha\beta}^{-+}$ and $(r - K) \times (r - K)$ with elements $P_{\alpha\beta}^{--}$. Arrange these submatrices into a matrix P_* where

$$P_* = \begin{pmatrix} P^{++} & P^{+-} \\ P^{-+} & P^{--} \end{pmatrix} \qquad 6.29$$

Now given an arbitrary configuration α, one may calculate the distance the tracer moves in one jump by the product

$$P'B(I - P_*)1 \qquad 6.30$$

where B is again a diagonal matrix with elements x_α, $(I - P_*)$ is the probability that the second jump does not occur. The I is the r-dimensional unit matrix, 1 is the N-dimensional unit column vector, and P' is also dependent on the applied field.

The distance for the second tracer jump follows as

$$P'P_*B(I - P_*)1 \qquad 6.31$$

This leads to a net distance for the second jump of the distance traveled on the first jump times the probability the second jump occurs, plus the distance on the second jump. Hence for a two jump sequence, one has a total distance of

$$P'B(I - P_*)1 + P'BP_*(I - P_*)1 + P'P_*B(I - P_*)1 \qquad 6.32$$

For an arbitrary sequence of consecutive jumps the total tracer displacement vector, S, is

$$S = P'\{[B + BP_* + BP_*^2 + \cdots] + [P_*B + P_*BP_* + P_*BP_*^2 + \cdots]$$
$$+ [P_*^2B + P_*^2BP_* + \cdot]\cdots\}(I - P_*)1 \quad 6.33$$

On carrying out the multiplication of Eq. 6.33 indicated by braces, one finds a simplification as follows:

$$S = P'(I - P_*)^{-1}B1$$
$$= P'(I - P_*)^{-1}d \qquad 6.34$$

where $d = B1$.

The elements in the vector S, $S\alpha$, are the distance the tracer moves in an infinite sequence of jumps starting with the configuration that follows a jump of type α.

The defect starts its motion at a source and wanders up to the tracer, causes the displacement S, and finds a sink. The sources and sinks of the defect are located on the surface of a large sphere with the tracer located at the center. One may denote the diagonal matrix, ν, whose elements, ν_α, are the rate at which a defect leaves the source and arrives at configuration following an α type jump.

Then the net tracer velocity is given by v_F as follows:

$$v_F = 1^T \nu S \qquad 6.35$$

The subscript F denotes that any contribution from a diffusion coefficient gradient has been omitted. The contribution to the velocity from the diffusion coefficient gradient was shown in Chapter 2 not to contribute to the flux. As a consequence, the flux of the tracer may be generally written as

$$J = -D^* \frac{\partial c}{\partial x} + cv_F$$

Random Walk Calculation of P and ν

One may consider a region of the lattice containing the tracer and identify a set of tracer-defect configurations, i, for $i = 1, \cdots M$; $M \geq r$. The defect may move from one configuration to another with a probability A_{ij}. Here A_{ij} is the probability that a single defect jump from the class of configurations i will leave the defect in a particular configuration in class j. For purposes of generality of approach, the matrix A is the transpose of the previous. One deletes tracer transitions in this matrix. Even though tracer jumps are not allowed in A, it is important that all tracer jump configurations be represented in M. Furthermore, self-image configurations that could lead to an $a+$ or $a-$ tracer jump must be represented twice in both A and P; once for a positive and once for a negative jump. Also, the elements A_{ij} are sensitive to the direction of the jump in relation to the applied field. One defines a matrix R^* of dimension $r \times M$; the elements in $R^*\alpha i$ represent the initial probability distribution. Finally, define a matrix Q^* of dimension $M \times r$. The elements $Q^*_{j\alpha}$ are the probability that a configuration j will, on the next defect jump, cause a tracer jump of type α, and that $Q^*_{j\alpha}$ is sensitive to the direction of the applied field. Then,

$$P_* = R^*(I - A)^{-1}Q^*$$

and

$$P' = R^*Q^* \qquad\qquad 6.36$$

It is of interest here to order the elements in A so that the jump configurations appear first. One must correspondingly order the elements in R^* and Q^*. Then it is obvious that these matrices can have the form

$$R^* = (R, 0)$$

$$Q^* = \begin{pmatrix} Q \\ 0 \end{pmatrix}$$

where R is the $r \times r$ matrix and Q is also $r \times r$. There exists a matrix A_e of dimension $r \times r$ such that

$$P_* = R(I - A_e)^{-1}Q \qquad\qquad 6.37$$

(see Appendix C).

With Eq. 6.37, one finds

$$S = RQ[I - R(I - A_e)^{-1}Q]^{-1}d \qquad\qquad 6.38a$$

or

$$S = R^*Q^*[I - R^*(I - A)^{-1}Q^*]^{-1}d \qquad\qquad 6.38b$$

When there is a single force and jump distance, $b = b_\alpha$ all α, as in the case of electromigration perpendicular to (100) planes of a cubic crystal, Eq. 6.38 will provide the more convenient description for most models because different defect jump rates will normally be considered as having become uniform by the time one has considered the r original defect configurations. In that instance, classes of defect jumps outside this r will only require knowledge of the average defect migration distance to calculate S.

In general, however, this simplification will not be present, and Eq. 6.38 will need to be used. The example is thermal diffusion and the vacancy wind effects. One must recognize that the purpose herein is to determine a quantity, such as the effective charge for electromigration, and thus one will want Eq. 6.38 linear in the applied force.

There are several terms in Eq. 6.38 that are linear in the applied field. Thus for the applied field one may write $S = S_0 + \delta S$, where S_0 is independent of applied forces and

$$S_0 = R^* Q_0^* [I - R^*(I - A_0)^{-1} Q_0]^{-1} d \qquad 6.39$$

where A_0 and Q_0^* are independent of the field. One can write the field effects as $A = A_0 + \delta A$ and $Q^* = Q_0^* + \delta Q^*$

$$\delta S = R^* \, \delta Q^* [I - R^*(I - A_0)^{-1} Q_0]^{-1} d$$
$$+ R^* Q_0^* \{ [I - R^*(I - A_0)^{-1} Q_0]^{-1} [R^*(I - A_0)^{-1} \, \delta Q$$
$$+ R^*(I - A_0)^{-1} \, \delta A (I - A_0)^{-1} Q_0] [I - R^*(I - A_0)^{-1} Q_0]^{-1} \} \, d \qquad 6.40a$$

which is of the form

$$\delta S = [\delta P'(I - P_0)^{-1} + P'(I - P_0)^{-1} \, \delta P(I - P_0)^{-1}] \, d \qquad 6.40b$$

Since from symmetry there are just as many positive as negative jumps from a given configuration, S_0 in Eq. 6.39 does not contribute to the flux. Consequently, Eq. 6.35 should have been written

$$v_F = 1^T \, \delta \nu S_0 + 1^T \nu_0 \, \delta S$$

with δS as given above, and $\nu = \nu_0 + \delta \nu$ as described below, in Eq. 6.43.

The general calculation of ν_π is as described previously. By contrast, to find ν_α, and thereby ν, one generally must consider the total defect path from the source up to the jump configuration, following α with possible arrival at all other jump configurations but no tracer jumps.

Define $T(\gamma)$ as the number of defects per unit time jumping from a set of outer defect source sites one jump removed from the configurations considered in A. It is linear in **an** applied field, and is written as M-component row vector T. Now ν is a row vector whose elements, ν_α, are the arrival rate of defects at sites following an α tracer type jump without the occurrence of tracer jumps, and receiving defects from all outer defect source sites. The probability that a defect path will reach the configuration following an α type jump from γ is $Y_{\gamma\alpha}$, so that

$$\nu = TY \qquad 6.41$$

where Y is an $M \times r$ matrix with elements $Y_{\gamma\alpha}$. The product TY is not generally diagonal, and the off-diagonal elements must be set to zero. If

H is a column vector whose elements, H_α, contain the number of configurations in class α, then the elements of the matrix Y are found in

$$Y_{\gamma\alpha} = [(I - A)^{-1}H]_{\gamma\alpha}$$

where A is sensitive to the field as defined before. Hence

$$\nu_\alpha = [T(I - A)^{-1}H]_{\alpha\alpha} \qquad\qquad 6.42$$

The use of the matrix A assures that all defect configurations are reached but no tracer jumps are included. As a result of Eq. 6.42, one can write

$$\begin{aligned}
\nu_\alpha &= \nu_{\alpha 0} + \delta\nu_\alpha \\
&= [T_0(I - A_0)^{-1}H + \delta T(I - A_0)^{-1}H + T_0(I - A_0)^{-1}\, \delta A(I - A_0)^{-1}H]_{\alpha\alpha}
\end{aligned}$$

$$6.43$$

where the last two terms are linear in the applied field.

TRACER MOBILITY BY A DIVACANCY MECHANISM

With the previous equations developed in this chapter, it is possible to approach the problem of diffusion by a divacancy mechanism where the divacancy is able to dissociate in the vicinity of the tracer. If the divacancy is unable to dissociate, one may use the formalism described in the previous chapter.

When the divacancy can dissociate at the tracer, four of the tracer jump configurations in fcc are screened from the source by other jump configurations. As a consequence $\nu_{\pi\alpha}$ used here and in Chapter 5 for these configurations, is zero, and this leads to both an erroneous diffusivity and tracer mobility. On the other hand, ν_α as calculated in this chapter avoids this problem; hence the approach of this chapter is necessary.

To approach this problem, one must realize that there are 18 divacancy configurations in fcc for a $\langle 100 \rangle$ diffusion direction where at least one of the vacancies neighbors the tracer. There are an additional 9 configurations where both vacancies neighbor the tracer but the vacancies are separated by at least one jump. This gives a total of 27 two-vacancy configurations. Also, if one of these configurations were to decompose so that only one vacancy neighbors the tracer and a divacancy does not exist, one may assume that the second-neighbor vacancy has become random. Then, there are three additional single vacancy-tracer configurations possible for a total of 30 as a minimum number of configurations in this problem. Table 6.1 identifies these configurations.

Table 6.1 Divacancy configuration classes[a]

Configuration number	Tracer at $(0, 0, 0)$		Number of configurations per class
	Vacancy 1	Vacancy 2	
1	$(1, 0, 1)$	$(0, -1, 1)$	8
2	$(1, 0, 1)$	$(1, -1, 0)$	4
3	$(2, 0, 0)$	$(1, -1, 0)$	4
4	$(1, 1, 0)$	$(0, 2, 0)$	4
5	$(1, 1, 0)$	$(2, 2, 0)$	4
6	$(1, 1, 0)$	$(2, 1, 1)$	8
7	$(1, 1, 0)$	$(1, 2, 1)$	8
8	$(0, 1, 1)$	$(1, 2, 1)$	8
9	$(1, 1, 0)$	$(1, -1, 0)$	2
10	$(1, 1, 0)$	$(0, -1, 1)$	8
11	$(1, 1, 0)$	—	4
$11 + I$, $I \leq 11$ is mirror image of I			
23	$(1, 1, 0)$	$(-1, -1, 0)$	4
24	$(1, 1, 0)$	$(-1, 0, 1)$	8
25	$(1, 1, 0)$	$(-1, 1, 0)$	4
26	$(0, 1, 1)$	$(0, -1, 1)$	4
27	$(0, 1, 1)$	$(0, -1, -1)$	2
28	$(0, 1, 1)$	$(0, 0, 2)$	8
29	$(0, 1, 1)$	$(0, 2, 2)$	4
30	$(0, 1, 1)$	—	4

[a] J. P. Stark, *J. Appl. Phys.* **46**, 2889 (1975).

Any physical model for the jumps of a divacancy at a solute tracer should include a reasonable number of variables. If one were to use the set of jump frequencies in the last chapter, one would find so many parameters that computation is impossible. Consequently, the simplifying model used previously is programmed. In this model x is the probability of a vacancy jump away from the tracer, y is the probability of a vacancy jump away from a neighboring vacancy, z is the increase ratio of the jump rate of a bound divacancy into the tracer, w_2 is the vacancy exchange rate with the tracer, and z_0 is the increase jump frequency ratio of the vacancy into all ions when either a divacancy is present or both vacancies neighbor on the tracer. The jump rate in the last circumstance is identified as w_0. If one identifies $q_i = q^*_i - \Delta H_{vi}$, as was done in Eqs. 5.60 and 5.61, then one has variables for the heat of transport of q_x, q_y, q_0, and q_2. Thus as in Eqs. 5.60 and 5.61, the effective heat of transport could possibly be dependent on all the previously mentioned variables. The matrix A is 30×30, as defined in Table 6.1. The matrix Q is 30×26, and R is 26×30. The dimension 26 comes from configurations 1–7, 9–11, 12–18, 20–22, 23, 24, and 25. The first 20 configurations have jumps in only one direction. The last three can give tracer jumps in both the positive and negative directions and hence must be included twice in both Q and R.

If V is the concentration of vacancies in the alloy, then it is apparent from the above model that the heat of transport may be represented as follows when w_0 is set to unity:

$$Q_{\text{eff}} = q_2(1 + z_0 V \alpha_1) + \frac{3x}{2+7x} q_x(1 + z_0 V \alpha_2)$$

$$-\frac{(4-3x)}{2+7x} q_0(1 + z_0 V \alpha_3) + z_0 V \alpha_4 q_y \qquad 6.44$$

It is implicitly assumed that the vacancy concentration is sufficiently low that the second vacancy makes only minor alteration in the heat of transport. Thus the divacancy correction is proportional to V, the vacancy fraction. Also, z_0 is present in all jump frequencies for the two vacancy configurations mentioned in the table, and it is therefore also found in Eq. 6.44. Based on this, the α_i's, $i = 1$–4, will be functions of z, x, y, and possibly w_2.

Numerically it was found that Q_{eff} was totally independent of w_2. Second, it was found that α_4 was about 0.1 or less for all cases studied. Thus the heat of transport for divacancy separation played no role in the final results. Furthermore, for $1 \leq y \leq 0.1$, all values of α_1, α_2, and α_3 were about 0.1 or less. This is consistent with α_4 being negligible. In fact, α_1–α_3 were only sizable when $y \to 10^{-2}$ or less. For such a small value of y, the vacancies are almost totally bound and a mixture of tracer jumps is found by isolated single vacancies and by bound divacancies.

The effective heat of transport for single vacancies was given in Eq. 5.60. For bound divacancies one has a comparable result, which was found numerically by letting $y \to 0$ in the program.

For the case of bound divacancies with this model, one may write the diffusivity as†

$$D = 16b^2 z w_2 N_{2\Gamma} f \qquad 6.45$$

when $N_{2\Gamma}$ is the divacancy concentration at the tracer, and b is the x projection of the jump distance in fcc. The form of Eq. 6.45 was

---

† This form is suggested by J. R. Manning, private communication.

confirmed numerically and the correlation factor may be written as

$$f = \frac{F^*x}{F^*x + z_0 z w_2/w_0} \tag{6.46}$$

where

$$F^* = \frac{1.09 + 0.99x}{1.00 + 0.60x} \tag{6.47}$$

One calculates the diffusivity using Eq. 6.15. The correlation factor is then found by taking the ratio of Eq. 6.7 to Eq. 6.7 with P_* set to zero. Thus the correlation factor may be written as

$$f = \frac{1^T v P'(I - P_*)^{-1} B(I + P_*)(I - P_*)^{-1} B1}{1^T v P'(I - P_*)^{-1} B^2 1} \tag{6.48}$$

for any arbitrary mechanism, even with multiple jump distances. Since this equation is general, it of course works for the bound divacancy as a limiting case of divacancies that dissociate.

The heat of transport for the bound divacancy mechanism may also be numerically fit to a closed functional form. The program indicates that one may write

$$Q_{\text{eff}} = q_2 + q_x \left\{ \frac{8.86x - 1.00}{8.82x + 2.91} \right\} + q_0 \left\{ \frac{1.50x - 1.00}{1.36x + 0.36} \right\} \tag{6.49}$$

The matrix equations developed in this chapter are completely general and one may, therefore, study diffusion in crystals for an arbitrary mechanism. The calculations in Chapters 3–5 are simplifications of these general results for cases where lattice symmetry permits the reduction of the number of configuration classes. Such symmetry should be used when possible because of the reduction in the size of the matrices to be manipulated numerically.

INFLUENCE OF A SOLUTE ON THE MOBILITY OF A SOLVENT TRACER

To study the manner in which a solute atom might influence the motion of a solvent tracer, use the same set of configurations as listed in the solute pair problem of the last chapter. Thus for an fcc crystal, one may take the configurations as listed in Table 5.9 as the starting point. In this case, however, the tracer is a solvent atom and the vacancy jump frequencies are the w_i, $i = 0, \cdots 4$, as defined in Chapters 3 and 4.

Table 6.2 Some of the nonzero matrix elements
Solute influence of solvent mobility[a]

$A(i, j)$

$A(1, 41) = w_4/B1$	$\delta A(1, 41) = w_4(q_4 + D1/B1)/B1$
$A(22, 20) = w_4/B1$	$\delta A(22, 20) = -\delta A(1, 41)$
$A(1, 39) = w_4/B1$	$\delta A(1, 39) = w_4(q_4 + D1/B1)/B1$
$A(22, 18) = w_4/B1$	$\delta A(22, 18) = -\delta A(1, 39)$
$A(1, 3) = w_0/B1$	$\delta A(1, 3) = w_0 D1/B1^2$
$A(22, 24) = w_0/B1$	$\delta A(22, 24) = -\delta A(1, 3)$
$A(2, 40) = w_0/B2$	$\delta A(2, 40) = w_0(q_0 + D2/B2)/B2$
$A(23, 19) = w_0/B2$	$\delta A(23, 19) = -\delta A(2, 40)$
$A(2, 3) = w_0/B2$	$\delta A(2, 3) = w_0 D2/B2^2$
$A(23, 24) = w_0/B2$	$\delta\ (23, 24) = -\delta A(2, 3)$
$A(3, 1) = 2w_0/B3$	$\delta A(3, 1) = 2w_0 D3/B3^2$
$A(24, 22) = 2w_0/B3$	$\delta A(24, 22) = -\delta A(3, 1)$
$A(3, 2) = 2w_0/B3$	$\delta A(3, 2) = 2w_0 D3/B3^2$
$A(24, 23) = 2w_0/B3$	$\delta A(24, 23) = -\delta A(3, 2)$

$Q(\beta)$

$Q_0(1, 1) = 4w_4/B1$	$\delta Q(1, 1) = 4w_4(q_4 + D1/B1)/B1$
$Q_0(2, 2) = 4w_4/B2$	$\delta Q(2, 2) = 4w_4(q_4 + D2/B2)/B2$
$Q_0(3, 3) = 8w_4/B3$	$\delta Q(3, 3) = 8w_4(q_4 + D3/B3)/B3$

$R(i, j)$

$R(1, 25) = \frac{1}{4}$	$R(2, 26) = \frac{1}{4}$
$R(3, 27) = \frac{1}{8}$	$R(4, 22) = \frac{1}{4}$

[a] $B1 = 4w_4 + 8w_0$
$D1 = 4(q_0 w_0 - q_4 w_4)$
$B2 = w_4 + 11w_0$
$D2 = q_0 w_0 - q_4 w_4$
$B3 = 2w_4 + 10w_0$
$D3 = 2(q_0 w_0 + q_4 w_4)$

In Table 5.9 there are 29 configurations in which the solute-vacancy-tracer are nearest neighbors of one another. These include 12 tracer jump configurations with nonzero x projections. To work this problem using the formalism of this chapter requires the 21 mirror-image configurations to those mentioned above. Also, there are 6 configurations in which the tracer-vacancy-solute are on the same x plane, and, three where the solute is not considered. These last three configurations are the ones in which the solvent tracer is to the left, right, and on the same x plane as the tracer. This gives a total of $21+21+6+3=51$ configurations. There are 12 tracer jump configurations in Table 5.9 where the tracer may make a positive jump of $+b$. The image of those 12 have x projections of $-b$. Also, there are the two configurations where only the vacancy and tracer are considered; one has a jump of $+b$ and the other $-b$. Hence there are $12+12+1+1=26$ tracer jump configurations in this problem.

Table 6.2 gives some of the jump probabilities for this problem. From these probabilities it is obvious that the effective heat of transport for thermal diffusion, $KT\mu/D = Q_e$ is a function of the frequency ratios w_4/w_0, w_2/w_1, and w_3/w_1. In addition, when the solute is not present one expects, from Chapter 4, that $Q_e = q_0/f_0$ where in the present approximation $f_0 = 9/11$. As a consequence, it is assumed that the heat of transport will follow

$$Q_e = q_0/f_0 + C \sum_{i=0}^{4} q_i \alpha_i (w_4/w_0, w_2/w_1, w_3/w_1) \qquad 6.50$$

Table 6.3 Results of solute influence on solvent mobility

w_3/w_1	α_0	α_1	α_2	α_3	α_4
			$w_4/w_0 = 1$		
0.01	−12.85	12.62	−6.25	0.44	0.58
0.10	−2.55	1.62	−0.49	0.42	0.57
1.00	−0.91	0.26	−0.00	0.28	0.44
10.00	−0.55	0.03	0.00	0.21	0.37
100.00	−0.50	0.00	0.00	0.19	0.36

w_4/w_0	α_0	α_1	α_2	α_3	α_4
			$w_3/w_1 = 1$		
0.01	−0.15	0.01	0.00	0.12	0.00
0.10	−0.22	0.03	0.00	0.13	0.05
10.00	−6.31	1.85	−0.27	1.13	2.79
100.00	−38.40	10.98	−2.67	5.34	16.11

where C is the solute concentration in units of atom fraction. The last portion, the summation, is the result of the influence of the solute on the solvent tracer. In numerical calculation it was found that a simplification was possible; namely, the functions α_i were independent of the frequency ratio w_2/w_1. Hence the $\alpha_i = \alpha_i(w_4/w_0, w_3/w_1)$. A brief tabulation of the parameters given in Eq. 6.50 is found in Table 6.3.

CHAPTER 7

SLOW AND FAST DIFFUSION
OF SUBSTITUTIONAL SOLUTES

DIFFUSION IN DISLOCATIONS AND GRAIN BOUNDARIES

It is now well-established that dislocations and grain boundaries provide crystalline heterogeneities in which diffusion can occur at a rate much faster than in the lattice. The small physical size of these defects, however, restrains the temperature range where the experimental effects are found; this is demonstrated in this following discussion.

For the sake of argument and presentation, we assume that diffusion of atoms that are normally substitutional solutes takes place by a vacancy mechanism even in the dislocation and grain boundary. As shown, there is some good evidence to substantiate this assumption. The principal kinetic factors controlling the diffusivity by a vacancy mechanism are the steady-state concentration of vacancies neighboring the solute as it resides in the dislocation or grain boundary, and the jump frequency of the solute into the vacancy. The correlation factor is also present, but that requires a separate analysis.

If one views the dislocation or grain boundary as a region where the atomic density is smaller than in the lattice, one would suspect that the vacancies have a positive binding energy to the defect. This implies that the concentration of vacancies at a solute atom is significantly larger at the defect, grain boundary, than in the lattice. Such an effect proportionately increases the solute diffusivity. Second, the decrease in density

195

should also manifest itself in an increased jump rate of the solute into the vacancy. With the exception of the correlation factor, this also proportionately increases the diffusivity. Based on previous discussion, an exceptionally large increase in the vacancy-solute jump frequency gives a proportionate decrease in the correlation factor. Nevertheless, because of the high concentration of vacancies at the solute and the increased vacancy-solute jump frequency, one expects a large increase in the solute diffusivity over that distance where the solute would be within the realm of influence of the dislocation or grain boundary.

The simplest model of a grain boundary from a diffusion standpoint is, therefore, a thin slab of width 2δ in which the solute diffusivity, D_b, may be orders of magnitude larger that that found in the lattice. This is the model first presented by Fisher[1] to explain the anomalous penetration found in polycrystalline diffusion samples. Since one is now viewing the sample as being composed of two regions of differing diffusivity, the lattice and the grain boundary, one must make appropriate modifications to Fick's second law since the mathematical solutions presented in Chapter 1 are invalid. One must derive alternate solutions that describe the heterogeneity through either the solution with multiple regions of differing diffusivity or appropriate boundary conditions. The conditions presented herein are idealizations designed to gain information on the diffusion coefficient in either a grain boundary or a dislocation.

The simplest experimental arrangement is for diffusion to be proceeding from either a constant or thin source at the plane $y = 0$. The defect, whether it is a single dislocation or a grain boundary, follows parallel to the y axis. In the case of a grain boundary, the center plane of the boundary would be located at $x = 0$. Matter can rapidly diffuse down the grain boundary with diffusivity D_b and half-width δ, then it can diffuse in the x direction out of the boundary and into the lattice at a diffusivity D_L. The grain boundary lattice interface is located along the plane $x = \pm\delta$.

The general solution to such a mathematical problem would be for diffusion to follow

$$\frac{\partial C_b}{\partial t} = D_b \, \nabla^2 C_b \qquad\qquad 7.1$$

when $x < \delta$; C_b is the tracer concentration in the boundary. When $x > \delta$, diffusion would follow

$$\frac{\partial C}{\partial t} = D_L \, \nabla^2 C \qquad\qquad 7.2$$

One would then insist that the conditions

$$C(\delta) = C_b(\delta)$$

and

$$D_L \frac{\partial C}{\partial x}\bigg|_{x=\delta} = D_b \frac{\partial C_b}{\partial x}\bigg|_{x=\delta} \qquad 7.3$$

be met to insure continuity of concentration and flux at the interface. One would apply initial conditions at the face $y = 0$. The solution to such a problem is a mathematical nightmare.

The situation is considerably simplified since δ is very small (a few Angstroms), and $D_b/D_L \gg 1$. This first implies from Eq. 7.3 that

$$\frac{\partial C_b}{\partial x} \ll \frac{\partial C_L}{\partial x}$$

at $x = \delta$. Second, since the boundary is centrally located at $x = 0$, $\partial C_b/\partial x$ is zero at $x = 0$ by symmetry. Those two conditions suggest that one could attempt a series representation of C_b in the variable x. Thus one assumes, as first shown by Whipple,[2] that

$$C_b(x, y, t) = C_{b0}(y, t) + \frac{x^2}{2} C_{b2}(y, t) \qquad 7.4$$

neglecting higher-order terms in the expansion. If one substitutes Eq. 7.4 into Eqs. 7.1 and 7.3, at $x = \delta$

$$D_b\left(\frac{\partial^2 C_{b0}}{\partial y^2} + C_{b2}\right) = \frac{\partial C_{b0}}{\partial t}$$

$$C_{b0} = C$$

$$\frac{D_b}{D_L} \frac{\partial C_{b2}}{\partial x} = \frac{\partial C}{\partial x} \qquad 7.5$$

when terms of order δ^2 have been neglected. The above may be combined into a single boundary condition to be used in the solution of Eq. 7.2, applicable at $x = \delta$. Combining this with Eq. 7.5, one finds that at $x = \delta$

$$D_b^* \frac{\partial^2 C}{\partial x^2} - \frac{D_L}{\delta} \frac{\partial C}{\partial x} = (\Delta - 1) \frac{\partial C}{\partial t} \qquad 7.6$$

where $\Delta = D_b/D_L$.

The mathematical problem, therefore, reduces to the solution of Eq. 7.2 with an initial or boundary condition at $y = 0$, and the condition in Eq. 7.6 represents the presence of the grain boundary. Whipple[3] solves this problem using a Fourier-Laplace transformation with the boundary condition that $C(x, y = 0, t) = 1$. Letting $\xi = (x - \delta)/\sqrt{D_L t}$, $\eta = y/\sqrt{D_L t}$, $\alpha = \delta/\sqrt{D_L t}$, and $\beta = (\Delta - 1)\alpha$, he finds

$$C = \operatorname{erfc} \frac{\eta}{2}$$

$$+ \frac{\eta}{2\sqrt{\pi}} \int_1^\Delta \frac{d\sigma}{\sigma^{3/2}} \exp - \frac{\eta^2}{4\sigma} \operatorname{erfc} \left[\frac{1}{2} \sqrt{\frac{(\Delta - 1)}{(\Delta - \sigma)}} \left(\xi + \frac{\sigma - 1}{\beta} \right) \right] \qquad 7.7$$

which is suitable for numerical analysis. For a given value of concentration, Eq. 7.7 would yield constant concentration contours as indicated in Fig. 7.1. The dotted line presents the limiting case in which no given boundary diffusion occurs, $\Delta \to 1$. Whipple also presents an approximate solution to the previous equation, he finds that for ξ small that

$$C = \operatorname{erfc} \frac{\eta}{2} + [1.159/(\eta\beta^{-1/2})^{3/2}]$$

$$\times \exp \{ -0.473(\eta\beta^{-1/2})^{4/3} + 0.396[(\eta\beta^{-1/2})^{2/3}/\beta]$$

$$\times (1 - \beta\xi) + 0[\beta^{-2}(1 - \beta\xi)^2] \} \qquad 7.8$$

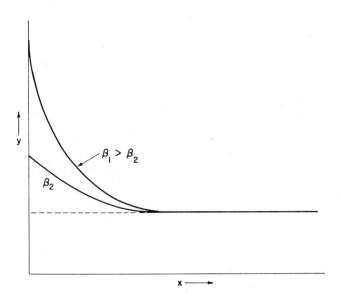

Figure 7.1   Isoconcentration contours for grain boundary diffusion.

which is appropriate near the boundary and, therefore, reflects the additional penetration of matter along the boundary. Equation 7.8 is particularly well-suited for the autoradiographic method of determining the grain boundary diffusivity. When serial sectioning is to be used, Suzuoka[4] has the preferable form to the thin film solution.

For diffusion along a dislocation "pipe" of high diffusivity that is placed centrally along the y axis, one can derive a similar set of equations to that found for the grain boundary slab. In such a circumstance, the concentration follows the differential equation[5]

$$D_L\left(\frac{\partial^2 C}{\partial r^2} + \frac{1}{r}\frac{\partial C}{\partial r} + \frac{\partial^2 C}{\partial y^2}\right) = \frac{\partial C}{\partial t} \qquad 7.9$$

The dislocation "pipe" lattice interface is at $r = a$, and one must satisfy a boundary condition similar to Eq. 7.6 at that place. Thus at $r = a$

$$D_P\frac{\partial^2 C}{\partial r^2} + \left(\frac{D_P - 2D_L}{a}\right)\frac{\partial C}{\partial r} = \left(\frac{D_P}{D_L} - 1\right)\frac{\partial C}{dt} \qquad 7.10$$

when D_P is the diffusion coefficient within the dislocation. Putting a constant concentration of $C = 1$ at the boundary $y = 0$, yields an approximate solution to Eqs. 7.9 and 7.10, as done by Stark. Thus

$$C \cong \frac{2}{\sqrt{\pi}} \frac{K_0\left[\frac{\beta\eta\xi}{2(1+\beta^2)}\right](1+\beta^2)\exp\{-\beta^2\eta^2/(4)(1+\beta^2)\}}{\beta\eta K_0\left[\frac{\beta\eta\alpha}{2(1+\beta^2)}\right]} \qquad 7.11$$

where

$$\xi = \frac{r}{\sqrt{(D_L t)}}$$

$$\eta = \frac{y}{\sqrt{(D_L t)}}$$

$$\alpha = \frac{a}{\sqrt{(D_L t)}}$$

and β must satisfy the additional constraint that

$$\beta^2 = \frac{K_2[\beta\eta\alpha/2(1+\beta^2)]}{(D_P/D_L - 1)K_0[\beta\eta\alpha/2(1+\beta^2)]} \qquad 7.12$$

where the K_0 and K_2 are Bessel functions.

The existence of the heterogeneity whether it originates with a grain boundary or a dislocation, significantly complicates the form of the mathematical solution to the diffusion equation. Nevertheless, experimentors have used such analyses to determine the rates of diffusion in both grain boundaries and dislocations.

Although it is probably not apparent from Eqs. 7.7, 7.8, or 7.11, the mass transport along the grain boundary or dislocation is very nearly parametric in the variable $\bar{P} = \delta D_b$. This may be seen through the approximate result of the grain boundary diffusion problem derived by Fisher.[6] He gets

$$C(x, y, t) = C_0 \exp \left\{ \frac{-y\sqrt{2}}{(\pi D_L t)^{1/4}(\delta D_b/D_L)^{1/2}} \right\} \operatorname{erfc}\left(\frac{x}{\sqrt{D_L t}}\right) \qquad 7.13$$

Fisher's solution, although not accurate enough for use with experimental data because it can lead to false conclusions, does correctly point out the parametric importance of the parameter $\bar{P} = \delta D_b$. The Fisher solution does actually give the correct interpretations when $\delta D_b/D_L$ is sufficiently large, and this circumstance was present in much of the early data on D_b.

The worthwhile experimental work on grain boundary diffusion has been motivated by the description of the boundary in terms of its structure. Low angle tilt boundaries are thought to be comprised of dislocations, and diffusion provides an experimental means of confirming such a belief. The first successful attempt in this direction was the work of Turnbull and Hoffman[7]. These researchers grew symmetric tilt bicrystals with misorientations of $9° \le \theta \le 28°$, where θ is the angle between adjacent (100) planes between the two crystals. Such bicrystals form grain boundaries comprised of dislocations with Burgers vectors b [100]; the dislocation density depends on the angle θ. If b is the unit cell dimension, the model of such a grain boundary would provide for a set of parallel dislocations with a spacing d; $b = 2d \sin \theta/2$. Turnbull and Hoffman surmise that the accelerated diffusion in the boundary is entirely contained within the dislocations themselves. If this is the case, then the diffusion parameter $\bar{P} = D_b \delta$, as found in the Fisher solution, should vary only with θ in its structural factor since the dislocation spacing is dependent on θ. Also, the diffusivity within the dislocation, D_P, should be independent of any variation in θ. Thus they presume that the diffusion parameter should follow

$$\bar{P} = D_b\, \delta(\theta) = D_P \frac{h^2}{d} = \frac{2h^2 D_P}{b} \sin \frac{\theta}{2} \qquad 7.14$$

where h^2 is the dislocation area in cross-section. The d is the dislocation separation distance from the Burgers model of a low angle tilt boundary; therefore, they determine $\bar{p}$ for values of θ between 9° and 28° for silver self-diffusion along grain boundaries using Fisher's mathematical solution. From their measurements they can plot $\ln \bar{p}$ versus $1/T$ and expect a series of straight lines parametric in θ. Indeed, this is what they found; for the values of $\theta = 9$, 13, and 16°, the experiments gave three parallel straight lines on a $\ln \bar{p}$ versus $1/T$ plot. For $\theta = 20°$, the slope of $\ln \bar{p}$ versus $1/T$ was indicative of a different activation energy, since for small values of θ, Eq. 7.14 should yield

$$\ln \bar{p} \propto - Q/K_B T \qquad\qquad 7.15$$

where $D_P = D_{P0}e^{-Q/K_B T}$ is the only activated parameter contained in $\bar{p}$. Hence the Q obtained was the activation energy for self-diffusion in dislocation pipes in silver with Burgers vectors of b [100]. Upthegrove and Sinnott performed the same type of analysis of self-diffusion in dislocations of nickel where the Burger's vectors would be b [100]; their results and others are presented in Table 7.1, including measurements for nickel with Burger's vectors of $b/2$ [110] obtained from grain boundary diffusion by Canon and Stark. Also, for comparison, Wüttig and Birnbaum, and Volin, Lie, and Balluffi have obtained data on diffusion along isolated dislocations; their results are also presented in the table. The former obtained their results by diffusing through thin slabs of deformed crystal, the latter through void shrinkage by dislocation diffusion currents.

Table 7.1   Activation energy for dislocation self-diffusion

System	Activation energy dislocation/lattice (e.v.)	Burgers vector	Reference
Ag	0.74/1.93	$b[100]$	a
Ni(edge)	1.76/2.95	$b/2[100]$	b
Ni(screw)	1.95/2.95	$b/2[110]$	b
Ni	1.08/2.95	$b[100]$	c
Ni	1.6/2.95	$b/2[110]$	d
Al	0.85/1.26	$b/2[110]$	e

[a] D. Turnbull and R. Hoffman, *Acta Met.* **2,** 419 (1954).
[b] R. F. Canon and J. P. Stark, *J. Appl. Phys.* **40,** 4361 (1969).
[c] W. R. Upthegrove and M. J. Sinnott, *Trans. ASM* **50,** 1031 (1958).
[d] M. Wüttig and H. K. Birnbaum, *Phys. Rev.* **147,** 495 (1966).
[e] T. Volin, K. Lie, and R. Balluffi, *Acta Met.* **19,** 263 (1971).

These results tend to emphasize that the activation energy is sensitive to the magnitude of the Burgers vector. Canon and Stark are able to justify the differences between their results and those of Upthegrove and Sinnott with a vacancy mechanism for diffusion. The vacancy-dislocation binding energy is sensitive to the square of the Burgers vector, and this is consistent with the magnitudes of energy between the two sets of data in nickel with varying Burgers vector.

The evidence presented above for an operative vacancy mechanism is sketchy at best. Nevertheless, a vacancy mechanism is consistent with the magnitudes of the diffusion coefficients and activation energies found experimentally. Such a mechanism would require a correlation factor to be present in the diffusion coefficient, and Robinson and Peterson[8] have investigated such a possibility both experimentally and theoretically. In what follows, we describe this development.

PURE METAL SELF-DIFFUSION—CORRELATION FACTOR FOR DISLOCATIONS

The model of Robinson and Peterson[8] views a dislocation with Burgers vector b [100] in an fcc crystal. They consider the tracer atom to be within the dislocation core in the crystal with the x direction parallel to the $\langle 100 \rangle$ and the central direction of the dislocation. They study the vacancy positions surrounding the tracer, which resides in the dislocation core as shown in Fig. 7.2. They also examine three groups of vacancy sites. Group I is the sites within the core, Group II is the sites Group I

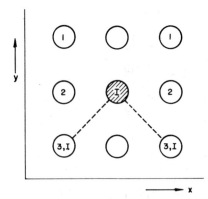

Figure 7.2   (100) Plane projection of dislocation (dotted line is core, tracer position is cross-hatched).

neighboring in the compressed region (those other sites viewed in Figures 7.2), and Group III is the second-nearest neighbor sites to the tracer. They construct a 12×12 matrix of vacancy random walk for each of the 12 neighbors to a tracer that sits at a particular Group I site (all vacancy sites are contained within Group II). We shall construct a similar model which uses the lattice symmetry to reduce the size of the 12×12 matrix. The jump frequencies for their model are described in Table 7.2.

Table 7.2 Dislocation-vacancy jump frequencies[9]

Initial group	Final group	Frequency
I	I	W_1
I	II	$W_2{}^1$
II	I	W_2
I	III	K^1
II	III	K
II	II	W_3

To reduce the 12×12 matrix using lattice symmetry, note that the center atom (the tracer) in Fig. 7.2 is symmetrical about the (100) plane through the tracer. Consequently, only the four vacancy sites to the right of the tracer, labelled 1, 2, and 3, must be used in the A matrix, which calculates f. Thus, as in Eq. 5.29, we calculate $P_{\alpha\beta}^a - P_{\alpha\beta}^{11}$ directly. This is possible because those vacancy paths that reach the x plane on which the tracer resides contribute equally to the antiparallel and parallel jump. These four vacancy sites to the right of the tracer can be reduced to three classes of vacancy sites. These classes are denoted 1, 2, and 3 in Figure 7.2; note there are two vacancy sites in Class 2 related by a mirror image through the z plane, which extends through the tracer, see ref. 9.

The correlation factor is calculated through the following equations

$$f_* = 1 + 2P_2(I - P_2)^{-1}1 \qquad\qquad 5.10$$
$$= 1 - 2T(I + T)^{-1}1$$
$$f = \underline{C}f_* \qquad\qquad 5.12$$

$$T = -P_2 = P_{\alpha\beta}^a - P_{\alpha\beta}^{11} = Q(\dot{\beta})(I - A)^{-1}R(\alpha^I) \qquad\qquad 5.29$$

Then 1 is the unit column vector, C is a row vector, and f_* is the vector whose components are the partial correlation factors.

It is convenient to define the following jump frequency combinations

$$d_1 = 2W_2 + 6W_3 + 4K$$

$$d_2 = 3W_2 + 5W_3 + 4K$$

$$d_3 = 2W_1 + 6W_2{}^1 + 4K^1$$

$$\delta_1 = \frac{W_3}{W_2}$$

$$\delta_2 = \frac{K}{W_2}$$

$$\delta_3 = \frac{W_2{}^1}{W_1}$$

$$\delta_4 = \frac{K^1}{W_1{}^1}$$

$$\delta_5 = \frac{W_1}{W_2} \qquad\qquad 7.16$$

where the last five relations were defined by Robinson and Peterson.

With these definitions, one can write the matrices of interest by inspection. Thus

$$A = \begin{pmatrix} 0 & \dfrac{2W_3}{d_2} & 0 \\[2ex] \dfrac{W_3}{d_1} & 0 & \dfrac{W_2{}^1}{d_3} \\[2ex] 0 & \dfrac{2W_2}{d_2} & 0 \end{pmatrix}$$

$$\underline{C} = \left(\frac{I_{II}}{3 + \delta_5}, \frac{2I_{II}}{3 + \delta_5}, \frac{\delta_5 I_I}{3 + \delta_5} \right) \qquad\qquad 7.17$$

where I_I and I_{II} are the concentration of vacancies on Groups I and II. One may also order the vectors Q and R into matrices so that the transpose of T is found from

$$T^T = Q(I - A)^{-1} R \qquad\qquad 7.18$$

where

$$Q = \begin{pmatrix} \dfrac{W_2}{d_1} & 0 & 0 \\[2ex] 0 & \dfrac{2W_2}{d_2} & 0 \\[2ex] 0 & 0 & \dfrac{W_1}{d_3} \end{pmatrix}$$

$$R = \begin{pmatrix} 0 & 0 & 1 \\ 0 & \frac{1}{2} & 0 \\ 1 & 0 & 0 \end{pmatrix} \qquad 7.19$$

This matrix form for T is very convenient since both T and A are of the same dimension, 3×3. We may take advantage of that fact, as shown in the following exercise.

$$f = 1 - 2T(1 + T)^{-1}$$
$$T^T = Q(I - A)^{-1}R$$
$$T = R^T(I - A^T)^{-1}Q^T$$
$$= R(I - A^T)^{-1}Q$$

since R and Q are symmetric. The function $T(I + T)^{-1}$ is desired in f. Hence we attempt to calculate it directly.

$$T(I + T)^{-1} = R(I - A^T)^{-1}Q[I + R(I - A^T)^{-1}Q]^{-1}$$
$$= \{[I + R(I - A^T)Q]^{-1}Q^{-1}(I - A^T)R^{-1}\}^{-1}$$
$$= \{Q^{-1}R^{-1} - Q^{-1}A^TR^{-1} + I\}^{-1}$$
$$= \{Q^{-1}(I - A^T + QR)R^{-1}\}^{-1}$$
$$= R(I - A^T + QR)^{-1}Q$$
$$= R(I - A_4)^{-1}Q \qquad 7.20$$

when

$$A_4 = A^T - QR \qquad 7.21$$

Thus one can write the correlation factor directly in terms of the new matrix A_4 as

$$f = 1 - CR(I - A_4)^{-1}Q1 \qquad 7.22$$

where

$$C = \frac{1}{3\delta_3 + 1}(\delta_3, 2\delta_3, 1)$$

and results from the average frequency

$$\langle W \rangle = 3W_2 I_{II} + W_1 I_I$$

with I_i as the concentration of vacancies in Group I. Thus if one lets $I_I = 1$, the definition of the frequencies in Table 7.2 implies that $I_{II} = \delta_3 \delta_5$; hence the values given for C_α. Also,

$$Q = \begin{pmatrix} \dfrac{1}{z} \\[2mm] \dfrac{2}{Q_0} \\[2mm] \dfrac{1}{u} \end{pmatrix}$$

$$A_4 = \begin{pmatrix} 0 & \dfrac{1}{y} & \dfrac{1}{z} \\[3mm] \dfrac{2}{\underline{x}} & -\dfrac{1}{Q_0} & \dfrac{2}{Q_0} \\[3mm] \dfrac{1}{u} & \dfrac{1}{w} & 0 \end{pmatrix} \qquad\qquad 7.23$$

where

$$Q_0 = 3 + 5\delta_1 + 4\delta_2$$

$$u = 2 + 6\delta_3 + 4\delta_4$$

$$w = 6 + \frac{2}{\delta_3} + \frac{4\delta_4}{\delta_3}$$

$$\bar{X} = 5 + \frac{3}{\delta_1} + \frac{4\delta_2}{\delta_1},$$

$$Y = 6 + \frac{2}{\delta_1} + \frac{4\delta_2}{\delta_1}$$

and

$$z = 2 + 6\delta_1 + 4\delta_2$$

The jump frequency ratios may now be converted into binding energies of the vacancy to the dislocation. These are given by

$$\frac{W_i}{W_j} = \delta_k = \exp - \frac{(\Delta E_i - \Delta E_j)}{k_B T} \qquad\qquad 7.24$$

where it is assumed that the pre-exponential vibrational frequencies ν_i must equal ν_j. One may construct a model whereby the vacancy is bound to the dislocation with a given binding for the jumps from Group I to Group II, Group II to Group II, and so forth. Such a model will then give the correlation factor based on the degree to which the vacancy is bound to the dislocation. As an example, if the dislocation were not present, $\delta_i = 1$ and Eqs. 7.22 and 7.23 yield the expected result that $f = \frac{9}{11}$. On the other hand, if we let $W_1 = 1$, $W_2^{\,1} = 0$, $W_2 = \infty$, $K^1 = 0$, and $W_3 = 1$ so that all δ are zero, $f = \frac{1}{3}$, which is not the expected result; one would find zero a more pleasing answer.

One problem that has arisen is that there is a difficulty in the values of $R(\alpha)$ as defined for jumps from Classes 1 and 2. The term $R(\alpha)$ is the configuration following a jump of type α. Following a jump of either type, Class 1 or 2 leaves the vacancy in a Group I site and the tracer in a Group II site. The previous set of sites only have the tracer in Group I sites. Expansion of the A matrix will alleviate the problem.

Define Class 4 to be the site following a jump of type 1, Class 5 to be the configuration following a jump of type 2, and Class 6 to be that following Class 3, which is the mirror image of Class 3. With these definitions the average jump frequency becomes

$$\langle W \rangle = W_2 I_{\text{II}} + 2 W_2 I_{\text{II}} + W_1 I_{\text{I}} + W_2^{\,1} I_{\text{I}} + 2 W_2^{\,1} I_{\text{I}} + W_1 I_{\text{I}}$$

$$= 3 W_2 \delta_3 \delta_5 + 2 W_1 + 3 W_2^{\,1}$$

leaving the C vector as

$$C = \frac{1}{6\delta_3 + 2} (\delta_3, 2\delta_3, 1, \delta_3, 2\delta_3, 1) \qquad 7.25$$

The R matrix is

$$R = \begin{pmatrix} 0 & 0 & 0 & 1 & 0 & 0 \\ 0 & 0 & 0 & 0 & \frac{1}{2} & 0 \\ 0 & 0 & 0 & 0 & 0 & 1 \\ 1 & 0 & 0 & 0 & 0 & 0 \\ 0 & \frac{1}{2} & 0 & 0 & 0 & 0 \\ 0 & 0 & 1 & 0 & 0 & 0 \end{pmatrix} \qquad 7.26$$

and Q is

$$Q = \begin{pmatrix} \dfrac{1}{z} & 0 & 0 & 0 & 0 & 0 \\[2mm] 0 & \dfrac{2}{Q_0} & 0 & 0 & 0 & 0 \\[2mm] 0 & 0 & \dfrac{1}{u} & 0 & 0 & 0 \\[2mm] 0 & 0 & 0 & \dfrac{\delta_3}{u} & 0 & 0 \\[2mm] 0 & 0 & 0 & 0 & \dfrac{2\delta_3}{u} & 0 \\[2mm] 0 & 0 & 0 & 0 & 0 & \dfrac{1}{u} \end{pmatrix} \qquad 7.27$$

Using these definitions of Q and R, and the old A matrix as a submatrix of this new 6×6 matrix, one combines Eqs. 7.26, 7.27, and the A submatrix to find

$$I - A_4 = \begin{pmatrix} 1 & -\dfrac{1}{y} & 0 & \dfrac{1}{z} & 0 & 0 \\[2mm] -\dfrac{2}{x} & 1 & -\dfrac{2}{Q_0} & 0 & \dfrac{1}{Q_0} & 0 \\[2mm] 0 & -\dfrac{1}{w} & 1 & 0 & 0 & \dfrac{1}{u} \\[2mm] \dfrac{\delta_3}{u} & 0 & 0 & 1 & 0 & 0 \\[2mm] 0 & \dfrac{\delta_3}{u} & 0 & 0 & 1 & 0 \\[2mm] 0 & 0 & \dfrac{1}{u} & 0 & 0 & 1 \end{pmatrix} \qquad 7.28$$

Then, Eqs. 7.25–7.28 can be used for numerical computation. The previous 3×3 analysis is of interest because one may make calculations by hand. The correlation factor is found from Eq. 7.22.

The above alters the temperature dependence of the correlation factor but does not change the lower bound of $f = \tfrac{1}{3}$. This value would reduce to the expected result of $f = 0$ if a large number of sites were included. The reason the correlation factor remains finite is that even when the vacancy is bound to the dislocation core, there is a large probability that the vacancy can randomize between tracer jumps by migrating outside the configurations incorporated in the A matrix. To remedy the situation, one may define the probability F of the vacancy escaping from the dislocation before returning to Class 3 or 6 sites by travelling along the dislocation core. Then the factor U, as used in the A_4 matrix, becomes $U = 2F + 6\delta_3 + 4\delta_4$. The situations $F = 1$ when $\delta_i = 1$ and $F = \frac{1}{2}$ when $\delta_i = 0$ for all i will reduce the correlation factor to the correct range. The only proper way to deal with the factor F is to use a much larger number of vacancy-tracer configurations. Nevertheless, a function that contains the proper limits on F is given by $F = 0.5 + 0.5\delta_3$. This function reflects the change in the vacancy binding to the dislocation core with temperature. Using this value for F along with $\delta_i = \exp(-E_i/K_B T)$, the correlation factor was calculated with the aid of Eqs. 7.22, 7.25–7.28 and the results are given in Table 7.3.

Table 7.3 Correlation factor comparison of one-and two-jump binding to the core[a]

T (°K)	$E_3 = E_4 = 3000$[b] (cal/mole)	$E_i = 3000$[c] (cal/mole)	$E_3 = E_4 = 2500$[b] (cal/mole)	$E_i = 2500$[c] (cal/mole)
500	0.29	0.25	0.41	0.35
550	0.35	0.30	0.47	0.41
600	0.41	0.35	0.52	0.45
650	0.46	0.40	0.56	0.49
700	0.50	0.44	0.59	0.53
750	0.54	0.48	0.62	0.56
800	0.57	0.51	0.65	0.58
850	0.60	0.53	0.67	0.60
900	0.62	0.56	0.68	0.62
950	0.64	0.58	0.70	0.64

[a] J. P. Stark, ref. 9.
[b] One-jump binding.
[c] Two-jump binding.

Robinson and Peterson[8] have measured $f \Delta K'$ using the isotope effect in silver. The $\Delta K'$ is the fraction of energy of the jumping tracer that remains at the saddle point. They find that $f \Delta K' \cong 0.45$ independent of temperature for the range of temperature from 623 to 823°K. The results of the calculation are given in Table 7.3 and Fig. 7.3. These results represent the situation in which vacancy jumps from Group I to Group II

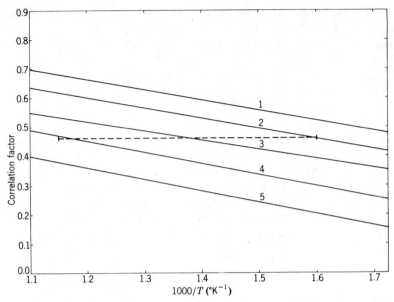

Figure 7.3 Correlation factor. (1) $E_i = 2000$ cal. (2) $E_i = 2500$ cal. (3) $E_i = 3000$ cal. (4) $E_i = 3500$ cal. (5) $E_i = 4000$ cal.

sites and from Group II to Group III sites are described by a binding potential of from 2 to 4 Kcal/mole. In the graph the calculation of f is presented; also, the measurements of $f \Delta K'$ is the horizontal line at $f \Delta K' = 0.45$. Since $\Delta K' \leq 1$, Cases III, IV, and V are unacceptable; f(calculated) $\geq f \Delta K'$ (measured). Hence the binding conditions of $E_i = 2$ and 2.5 Kcal/mole are the only conditions in agreement with the data.

Table 7.3 gives a comparison of one- and two-jump binding conditions. The difference is very small and suggests that little confidence may be placed in the jump energy from Group II to Group III sites.

It is notable that when $E_3 = E_4 = \infty$, $f = 0$ and, that when $E = 0$, $f = 0.83$ as expected.

INTERSTITIAL DIFFUSION OF NORMALLY SUBSTITUTIONAL SOLUTES

It is normally expected that the dissolution of one metal in another will result in a substitutional solid solution because the size of metal atoms are for the most part comparable. Thus the diffusivity of a metal solute in a metal solvent is expected, at least for close-packed lattices, to occur by a vacancy mechanism. The probability that a reasonable fraction of the solute is contained within the interstitices is thus negligible. Therefore, the diffusivity of the solute is expected to be roughly comparable with the self-diffusion of the solvent.

Such is not always the case, and a group of monovalent solutes dissolved in lead, tin, thallium, and indium diffuse at rates from two to eight orders of magnitude faster than the solvent tracers. Anthony, Dyson, and Turnbull[10] have found that the solutes silver, gold, and copper all diffuse at these ultrafast rates in the above mentioned solutes.

Based solely on the experimental observations, one must conclude that a significant fraction of the solutes reside in interstitial sites. Such a conclusion is understandable if the electrons from solutes interact with those of the solvent in some unusual way to depress the solute ionic diameter. Anthony and Turnbull[11] suggest that such a d–d shell interaction exists. Their model predicted that gold, silver, and copper would dissolve at least partially interstitially in those solvents. Furthermore, the interstitial/substitutional solute-size ratio would be higher for copper to silver to gold. Therefore, the diffusivities should follow that order; this was substantiated by experiment. Other monovalent metal solutes have also shown this ultrafast diffusion in lead, tin, thallium, and indium.

The diffusion coefficient for the mixture of interstitial-substitutional solutes, D_{eff} is related to the concentration of interstitial solute C_i and of substitutional solute C_s through

$$D_{eff} = \left(\frac{C_i}{C_i + C_s}\right) D_i + \left(\frac{C_s}{C_i + C_s}\right) D_s \qquad 7.29$$

where D_i is the diffusivity of the interstitial and D_s is for the substitutional. Since $D_i \gg D_s$ because D_s is expected to require a vacancy, $C_i/(C_i + C_s) < C_s/(C_i + C_s)$ and one might still observe $D_{eff} \gg D_s$. These are the conditions expected for the systems mentioned.

A more recent viewpoint of the mechanism that permits the interstitial diffusion of a normally substitutional solute is described by the following model as first suggested by Frank and Turnbull[12] and developed extensively by Miller[13] and Warburton.[14] In this model the fast diffusion results from a substitutional solute leaving its site at a rate V_0 to acquire an interstitial position thus creating a bound vacancy—interstitial pair in an fcc lattice. Now both defects diffuse rapidly even though they are bound, giving rise to the fast diffusion found experimentally. The bound vacancy can jump into a neighboring solvent at a rate V_1, the interstitial and vacancy can recombine at a rate K_2, and the interstitial can jump into a site adjacent to the vacancy and the interstitial original position at a rate

K_1. This leaves the diffusivity of the interstitial—vacancy pair as determined by McKee[15] to be

$$D_i = \frac{b^2}{3} \frac{(4K_1 + K_2)4V_1P}{4K_1 + K_2 + 4V_1}$$

and

$$P = 1/(1 + K_2/6V_0)$$

NEAR-SURFACE ANOMALY

Probably the most popular experimental condition for the determination of solute self-diffusion in a dilute alloy is that of a thin plane source of isotope deposited on a specimen of sufficient length to be considered semi-infinite. In such a case the solution to Fisk's second law is found to be

$$C(x, t) = \frac{Q}{\sqrt{(\pi Dt)}} \exp\left(-\frac{x^2}{4Dt}\right) \qquad 7.30$$

as given in Chapter 1. The diffusion coefficient is found by serial sectioning after a diffusion time t. One measures concentration versus x and finds D from the slope of a plot of $\ln c$ versus x^2.

In the past few years, results have been found that differ from the expected solution to Fick's laws as given in Eq. 7.30. Namely, the slope on a plot of $\ln c$ versus x^2 is found to have an anomalously steep slope very near the surface of the diffusion specimen, as indicated in Fig. 7.4 as Region I. In the second region (II) on that figure, the results are typical of those expected for lattice diffusion of substitutional solutes. If the crystal quality is poor, there would be a third region associated with dislocations acting as short circuit paths.

Evidently, something is slowing the diffusion of the solute isotope in the vicinity of the free surface of the diffusion specimen. One quantitative theory, presented herein, to explain this surface hold-up of the isotope has been presented by Reimann and Stark[16].

It had been shown previous to the Reimann and Stark work that the diffusion of silver and copper into lead demonstrated a near surface effect, NSE, due to the formation of lead oxide on the surface of the diffusion specimen. This is a case where the formation free energy of the solvent oxide is larger in magnitude that the solute. Thus once the solvent oxide is reduced, the solute is free to diffuse into the material. The opposite is true for the diffusion of cobalt into silver[17] and zinc into copper,[18] each of which had shown a near surface effect. Reimann and

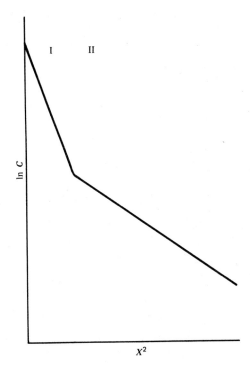

Figure 7.4 Isotope penetration with near surface effect present.

Stark propose that the more stable solute oxide is attempting to diffuse into the solvent, eventually breaks up due to thermal fluctuations, then diffuses at its characteristic rate.

Experimentally, the formation of oxides is almost never excluded; a hydrogen atmosphere, which is a common preventive means, only has a limited applicability. Also, many isotope tracers have oxides that are more stable than the H_2O molecule that would form with the hydrogen atmosphere. Thus, in many experimental situations, the thin layer of tracer isotope is in an initial state of being an oxide rather than a free metallic film on the specimen. This would be particularly true with the common vapor deposition technique. Consequently, the specimen is visualized as having a thin layer of tracer oxide on its surface at time $t = 0$. The tracer oxide may diffuse with a characteristic diffusion coefficient of D_0 and with a concentration $c_0(x, t)$. Once the oxide molecule breaks up, the free isotope of concentration C_i may diffuse with a diffusivity D_i. The initial conditions are therefore $C_0(x, t = 0) = Q_0 \delta(x)$, where $\delta(x)$ is the usual Dirac delta function and $C_i(x, t = 0) = 0$. Assuming the rate at which the oxide molecule dissociates is λ, one may write simultaneous differential equations for the motion of tracer oxide and free tracer, see Chapter 1. Thus the concentrations must satisfy

$$\frac{\partial C_0}{\partial t} = D_0 \frac{\partial^2 C_0}{\partial x^2} - \lambda C_0 \qquad 7.31a$$

$$\frac{\partial C_i}{\partial t} = D_i \frac{\partial^2 C_i}{\partial x^2} + \lambda C_0 \qquad 7.31b$$

where it is assumed that the concentration of oxygen in the specimen is sufficiently small that reformation of the oxide is negligible. If such a situation were not present one would have another term, $\lambda_2 C_i$, in each equation. In what follows we examine the solution to Eq. 7.31 and the theory for the value of λ, which gave quantitative agreement with the observations. The equations may be solved using Laplace transform techniques as outlined in Chapter 1. The method follows the assumption that

$$C_0 = f(t)g(x, t)$$

from which it follows that

$$C_0 = \frac{Q_0}{\sqrt{\pi D_0 t}} \exp\left\{ -\left(\lambda t + \frac{x^2}{4D_0 t} \right) \right\} \qquad 7.32$$

from Eq. 7.31a. Inserting Eq. 7.32 into 7.31, and taking the Laplace transform with p replacing the variable t, gives a substitute equation for 7.31 as

$$\frac{\partial^2 \bar{C}_i}{\partial x^2} = \frac{P}{D_i} \bar{C}_i - \frac{\lambda Q_0 \exp\left[-x\sqrt{(P+\lambda)/D_0} \right]}{D_i D_0 [(P+\lambda)/D_0]^{1/2}} \qquad 7.33$$

where

$$\bar{C}_i(x, P) = \int_0^\infty e^{-Pt} C_i(x, t)\, dt \qquad 7.34$$

One can solve Eq. 7.33 as

$$\bar{C}_i = A e^{-\sqrt{(P/D_i)}x} + B e^{-\sqrt{[(P+\lambda)/D_0]}x} \qquad 7.35$$

where

$$B = \frac{-\lambda Q_0}{D_0 D_i \sqrt{(P+\lambda)/D_0}\, [P(1/D_0 - 1/D_i) + \lambda/D_0]} \qquad 7.36$$

is found by substituting 7.35 into Eq. 7.33. The constant A is determined from the boundary condition

$$\int_0^\infty [C_0(x, t) + C_i(x, t)]\, dx = Q_0$$

which transforms into

$$\int_0^\infty [\bar{C}_0(x, P) + \bar{C}_i(x, P)]\, dx = \frac{Q_0}{P} \qquad 7.37$$

Then,

$$A = Q_0\sqrt{P/D_i}\left\{\frac{\lambda}{P(P+\lambda)} + \frac{1}{[P+\lambda][(P/\lambda)(\Delta-1)+\Delta]}\right\} \qquad 7.38$$

when $\Delta = D_i/D_0$.

Combining Eqs. 7.35, 7.36 and 7.38, and taking the inverse transform assuming $\Delta \gg 1$, one finds

$$C_i(x, t) = Q_0\lambda \int_0^t \frac{e^{-x^2/4D_i\Gamma} e^{-\lambda(t-\Gamma)}\, d\Gamma}{\sqrt{\pi D_i \Gamma}}$$

$$-Q_0\lambda e^{-\lambda t}\frac{D_0}{D_i}\int_0^t \frac{e^{-x^2/4D_0\Gamma}\, d\Gamma}{\sqrt{\pi_0\Gamma}}$$

$$+\frac{Q_0}{\sqrt{\pi}}\lambda D_0 \int_0^t (t-\Gamma)e^{-x^2/4D_i\Gamma}\left[\frac{4(D_i\Gamma)^{5/2}}{x^2} - \frac{1}{2(D_i\Gamma)^{3/2}}\right] d\Gamma$$

The solution is unwieldy; however, it may be integrated numerically knowing λ, D_0, D_i, t, and x for any value of C.

Given an observation of the NSE, one will have $C_0 + C_i$ as functions of x and t. From the sum of these, one may determine D_0 from a plot of $\ln(C_0 + C_i)$ versus x^2 near the surface, and D_i from the slope farther out in the specimen. Hence everything is known except λ, which must be determined either theoretically or experimentally. To validate this model for the near surface effect, theoretical understanding of λ is necessary.

Given that the free energy of formation of the solute tracer is ΔG_t and that of the solvent is ΔG_s, the net free energy change for the reaction that reduces the solute oxide is

$$\Delta G^* = \Delta G_t - \Delta G_s \qquad 7.40$$

Now as the tracer oxide sits in solution in the solvent, it is constantly bombarded with thermal energy and hence attempts to break apart. The probability per cycle that the oxide would break apart is

$$P_1 = \exp(-\Delta G^*/K_B T) \qquad 7.41$$

for each phonon reaching the molecule. Now for the molecule to break apart, one must move the solute tracer away from the oxygen ion. This may be understood when one considers the ionic diameters involved. The oxygen ion is negatively charged and hence very large in size. The $0^=$ ion is of a size comparable with most metallic neutral atoms. Thus one must view the oxygen as being in a substitutional site. The small metallic solute ion is possibly also in a substitutional position as the diffusivity indicates. To break apart the molecule, one must move the solute tracer ion to a neighbor substitutional site and at the same time break apart the molecule with probability given by Eq. 7.41. Thus the diffusion process occurs at the same time and one needs a second probability of

$$P_2 = \exp - \left(\frac{\Delta H_v + \Delta H_m}{K_B T} \right) = \exp - \left(\frac{\Delta H_D}{K_B T} \right) \qquad 7.42$$

where ΔH_v is the formation energy of a vacancy at the solute and ΔH_m is the migration energy. The rate constant, λ, is the number of forced oscillations per second, ν (the Debye frequency), times $P_1 P_2$. Hence we view

$$\lambda = \nu \exp \left\{ \frac{-(\Delta H_D + \Delta G^*)}{K_B T} \right\} \qquad 7.43$$

This value of λ along with the previous theory for oxide dissociation was capable of quantitatively explaining the data on zinc in copper, cobalt in silver, gallium self-diffusion, and cadmium in silver.[19] All these systems show the previous experimental behavior as indicated in Fig. 7.2.

At low temperatures the near surface anomaly apparently goes away. This can also be explained by oxidation kinetics.[19]

REFERENCES

1. J. C. Fisher, *J. Appl. Phys.* **22**, 74 (1951).

2. R. T. P. Whipple, *Phil. Mag.* **45**, 1225 (1954).

3. R. T. P. Whipple, ibid.

4. T. Suzuoka, *Trans. Jap. Inst. Met.* **2**, 25 (1961).

5. J. P. Stark, *J. Appl. Phys.* **36**, 3938 (1965).

6. J. C. Fisher, ibid.

7. D. Turnbull and R. Hoffmann, *Acta Met.* **2**, 419 (1954).

8. J. T. Robinson and N. L. Peterson, *Acta Met.* **21**, 1181 (1973); *Surface Sci.* **31**, 586 (1972).

9. J. P. Stark, *Acta Met.* **23**, 667 (1975).

10. T. R. Anthony, B. F. Dyson, and D. Turnbull, *J. Appl. Phys.* **37,** 2925, 2370 (1966).

11. T. R. Anthony and D. Turnbull, *Appl. Phys. Lett.* **8,** 120 (1966).

12. F. C. Frank and D. Turnbull, *Phys. Rev.* **104,** 617 (1966).

13. J. W. Miller, *Phys. Rev.* **108,** 1074 (1969).

14. W. K. Warburton, *Phys. Rev.* **B7,** 1341 (1973).

15. R. A. McKee, *Phys. Rev.* **B13,** 635 (1976).

16. D. K. Reimann and J. P. Stark, *Acta Met.* **18,** 63 (1970).

17. T. S. Lundy and R. A. Padgett. *Trans. AIME* **242,** 1897 (1968).

18. D. L. Styrus and C. T. Tomizyka, *J. Appl. Phys.* **34,** 1001 (1963).

19. C. M. Teller and J. P. Stark, *Acta Met.* **20,** 1077 (1972).

STATISTICAL MECHANICS OF VACANCY-DIVACANCY EQUILIBRIUM IN A PURE fcc METAL

In a manner similar to that shown in Chapter 2, one may formulate the concentration of divacancies using statistical thermodynamics. This calculation requires a more difficult partition function than was used in Chapter 3 to show the existence of vacancies. Namely, one must enumerate the configurational degeneracy of a system of atoms in which the energy of a neighboring pair of vacancies differs from the two isolated point defects.

To accomplish this we select an fcc metal and introduce n total point defects in a system of N atoms, giving $N + n$ total lattice sites. It is assumed at the outset that $N \gg n$. We split these vacancies into two groups; group I has m isolated single vacancies, and group II has K divacancies. Hence $n = m + 2K$. We identify the formation energy of defects in groups I and II as E_1 and E_2, respectively. Thus the total energy of the system, excluding the cohesive energy of the system without vacancies; is E_T where

$$E_T = mE_1 + KE_2 \qquad \text{A.1}$$
$$= nE_1 + K(E_2 - 2E_1)$$
$$= nE_1 - K \, \Delta H_b \qquad \text{A.2}$$

The ΔH_b is the binding energy that holds the vacancies together.

We write the partition function as Q where

$$Q = \sum_{n,k} \Omega_1 \Omega_2 \exp\left(-E_T/K_B T\right) \qquad \text{A.3}$$

$$Q = \sum F(n, K)$$

where Ω_1 is the number of ways of introducing the single isolated vacancies into the lattice without producing a divacancy, and Ω_2 is the remaining ways of putting in the divacancies. Using arguments that are identically analogous to those that determine the chemical potential of the interstitial solute in Chapter 2 by using statistical mechanics, one would determine

$$\Omega_2 = \frac{(N+n)(N+n-12)\cdots[N+n-(m-1)12]}{m!}$$

$$= \frac{12^{n-2K}[(N+n)/12]!}{m!\,[(N+n)/12-m]!} \qquad \text{A.4}$$

and

$$\Omega_2 = \frac{\begin{array}{c}(N-11n+24K)(12)(N-11n+24K-20)(12)\cdots\\ [N-11n+24K-(K-1)20]\end{array}}{2^K K!}$$

$$= \frac{[(N-11n+24K)/20]!(20)^K(12)^K}{2^K K!\,[(N-11n+24K)/20-K]!} \qquad \text{A.5}$$

One may use the maximum term method on Eq. A.3, where Eqs. A.1, A.2, A.4, and A.5 are used to expedite the analysis. Thus one would find from $\partial \ln F/\partial K$ and $\partial \ln F/\partial n$ a pair of equations that yield the closest approximation to thermodynamics equilibrium. One finds

$$\left.\frac{\partial \ln F}{\partial K}\right|_n = \frac{\Delta H_b}{K_B T} - 2 \ln 12 + \ln 6$$

$$+ 2 \ln (n-2K) - \ln K - 2 \ln [(N+n)/12 - n + 2k]$$

$$+ \ln [(N-11n+24K)/20 - K] + \ln 20 = 0 \qquad \text{A.6}$$

$$\left.\frac{\partial \ln F}{\partial n}\right|_K = \frac{-E_1}{K_B T} + \ln 12 - \ln (n-2K)$$

$$+ \ln [(N+n)/12 - n + 2K] = 0 \qquad \text{A.7}$$

Using these equations (A.6 and A.7) with the additional approximation that $n \gg 2K$, $N \gg n$, one finds from Eq. A.7 that

$$n/N = \exp-(E_1/K_B T) \qquad \text{A.8}$$

as expected. In addition the number of divacancies is found from A.6 to be the expected result:

$$\frac{K}{N} = 6\left(\frac{n}{N}\right)^2 \exp\left(\Delta H_b/K_B T\right) \qquad \text{A.9}$$

Equation A.9 is derived as supplementary evidence to the kinetic analysis for the existence of divacancies from the theoretical viewpoint. The experimental evidence is now quite pronounced, as discussed in the text. For example, the Peterson, Rothman, and Robinson results are the correlation factor for self-diffusion of silver, and show that half the tracer jumps are by divacancies at the melting point. Additional information has been found by Balluffi and coworkers for the direct experimental observation of divacancies using field immersion microscopy. Also, Seeger and coworkers have fit much experimental diffusion data for pure metals to a mixed vacancy-divacancy mechanism; this leads to a curved plot of ln D versus $1/T$, which has been observed, particularly in bcc metals.

APPENDIX B

QUANTUM THEORY FOR DIFFUSIVE JUMPS[1]

In Chapter 2 a semiclassical theory of diffusive jumps was presented, in which the jump frequency was found to be exponentially dependent on temperature. Therein we found that a satisfactory explanation of the jump frequency related the vibrational frequency and the probability that a fluctuation occurred, permitting the atom to jump at its characteristic rate. Such a viewpoint is valid at high temperatures when the actual temperature $T \gg \theta_D$, the Debye temperature. In that circumstance the same temperature dependence of the jump frequency is found in a strict quantum mechanical calculation and represents the probability that the atoms comprising the saddle point move so as to permit the atomic displacement into the available and empty adjacent site.

At low temperatures for light nuclei, such as a proton, the probability of the diffusing atom jumping to the adjacent site without the exponential probability becomes comparable to the fluctuation probability. At that point the temperature dependence of the jump frequency between site p and site p', $\Gamma_{pp'}$, becomes proportional to the temperature to the seventh power. Thus at high temperatures,

$$\Gamma_{pp'} \propto \exp - (E/K_B T) \qquad \text{B.1}$$

which at low temperatures becomes

$$\Gamma_{pp'} \propto (T/\theta_D)^7 \qquad \text{B.2}$$

with a smooth continuous transition in between. To understand the manner in which this is found, the following discussion parallels some of the presentation of Flynn and Stoneham[1].

We consider the system in this instance to be represented by the lattice plus one interstitial atom, which is to be the diffusing species. The generalization to other mechanisms is straightforward. We define a set of localized eigenfunctions represented by the interstitial position, p, and the vibrational spectrum, ν, to be given by $|p, \nu\rangle$. It is important to remember that such a state cannot represent the exact eigenfunctions because these reflect the translational symmetry of the lattice. Thus we expect that the exact eigenstates are linear combinations of these localized states. The exact eigenfunctions represented by such states diagonalize the Hamiltonian, H, of the entire system. If we instead use these localized and approximate eigenfunctions, the Hamiltonian of the system is not diagonal. Furthermore, the expectation values of off-diagonal terms are related through first-order perturbation theory to the transition probability from one localized state, $|p, \nu\rangle$, to another localized state, $|p', \nu'\rangle$. Such a transition implies the motion of the interstitial atom to an adjacent interstitial as long as $|p', \nu'\rangle$ is another localized state. Thus the atom has moved from interstice p to interstice p'.

First-order perturbation theory gives the transition probability for this transition to be

$$w_{pp'}(\nu, \nu') = \frac{2\pi}{h} |\langle p\nu| H |p'\nu'\rangle|^2 \, \delta(E_{p\nu} - E_{p'\nu'}) \qquad \text{B.3}$$

where

$$E_{p\nu} = \langle p\nu| H |p\nu\rangle$$

In a diffusive jump one is concerned with the transition from one site to an adjacent site. Thus the particular state of the phonon spectrum is inconsequential; one needs to sum over all phonon states. As a result, one defines the overall jump rate as

$$W_{pp'} = \left\langle \sum_{\nu'} w_{pp'}(\nu, \nu') \right\rangle_\nu \qquad \text{B.4}$$

where the $\langle \rangle_\nu$ denotes the thermal average for the phonon spectrum. It is interesting that within Eq. B.4, no question even arises as to the details of atomic jump. No saddle point configuration is defined; only the initial and final state are important. Such details are not evident in the fluctuation theory approach for high temperature diffusion as discussed in Chapter 2. It appears that the saddle point picture is a consequence of absolute rate theory and an unnecessary restriction on the theoretical viewpoint.

To evaluate Eq. B.4 is a reasonably difficult job, even with simplifying approximations. The two major assumptions used by Flynn and Stoneham are that the interstitial ion is light, like a proton, and that the Born-Oppenheimer approximation is valid. In such an instance, the Hamiltonian, H, may be written as the sum of three terms as follows:

$$H = H_I + H_{int} + H_L \qquad \text{B.5}$$

where for an interstitial of mass m,

$$H_I = -\frac{h^2}{2m}\nabla^2 \qquad \text{B.6}$$

The potential of interaction between the lattice and interstitial is found in

$$H_{int} = \sum_i U(R - R_i, X_i) \qquad \text{B.7}$$

where R is the rest position of the interstitial, R_i is that of the host atom, and X_i is the displacement. Finally, H_L is the lattice energy represented by the phonons. Thus with the adiabatic approximation, the wave function of the entire system containing the interstitial at site p is found to be

$$\psi_p(R, X) = \tilde{\phi}(R, X)A_p(X) \qquad \text{B.8}$$

The interstitial wave function $\tilde{\phi}(R, X)$ satisfies

$$(H_I + H_{int})\tilde{\phi}(R, X) = E_p(X)\tilde{\phi}(R, X) \qquad \text{B.9}$$

whereas the lattice wave function satisfies

$$[H_L + E_p(X)]A_p(X) = EA_p(X) \qquad \text{B.10}$$

When E_p is expanded into a power series

$$E_p(X) = E_{p0} + A_p \cdot X \qquad \text{B.11}$$

it is possible to push through to an answer that yields B.1 and B.2 under the conditions described.

Interestingly, Baker and Birnbaum[2] have done low temperature measurements of hydrogen in niobium using internal friction. They find that the hydrogen jump rate satisfies Eq. B.2 at low temperatures, confirming the Flynn and Stoneham theory.

REFERENCES

1. C. P. Flynn and A. M. Stoneham, *Phys. Rev.* **B1**, 3966 (1970).
2. H. K. Birnbaum and C. Baker, *Ber Bun Ges* **76**, 827 (1972).

APPENDIX C

MATRIX ALGEBRA FOR A_e

One orders the matrix elements in A into classes of configurations as follows: class a contains the N configurations in which a tracer jump can occur; class b is one defect jump removed from class a, and so forth. The matrix A can then be arranged into the following form:

$$
A = \begin{pmatrix}
A(aa) & A(ab) & 0 & 0 & 0 \\
A(ba) & A(bb) & A(bc) & 0 & 0 \\
0 & A(cb) & A(cc) & A(cd) & 0 \\
0 & 0 & A(dc) & A(dd) & \cdot \\
0 & 0 & \cdot & \cdot & \cdot \\
\cdot & \cdot & & &
\end{pmatrix}
$$

By expanding the elements of $(1-A)^{-1}$, one may determine the effective matrix (A_e) of dimension $N \times N$ by the following procedure:

$$A_e = p_1(aa)$$
$$p_1(aa) = A(aa) + A(ab)P(bb)A(ba)$$
$$P(bb) = [1 - P_1(bb)]^{-1}$$
$$p_1(bb) = A(bb) + A(bc)P(cc)A(cb)$$
$$P(cc) = 1 - P_1(cc)^{-1}$$
$$P_1(cc) = A(cc) + A(cd)P(dd)A(dc)$$
$$P(dd) = \text{etc.}$$

the series may be cut off at any convenience boundary by setting $p(zz) = 0$.

APPENDIX D

OUTSIDE READING

GENERAL

1. C. P. Flynn, *Point Defects and Diffusion*, Clarendon, Oxford, 1972.
2. J. R. Manning, *Diffusion Kinetics for Atoms in Crystals*, Von Nostrand, Princeton, New Jersey, 1968.
3. P. G. Shewmon, *Diffusion in Solids*, McGraw-Hill, New York, 1962.
4. N. L. Peterson, "Diffusion in Metals" in *Solid State Physics*, Vol. 22 (F. Seitz, D. Turnbull, and H. Ehrenreich, Ed.), Academic, New York, 1968.
5. *Diffusion* American Society for Metals, Metals Park, 1973.

CHAPTER 1

1. S. Crank, *Mathematics of Diffusion*, Oxford, London, 1955.
2. J. Nye *Physical Properties of Crystals*, Oxford, London, 1957.

CHAPTER 2

1. L. Darken, *Trans. AIME* **180,** 430 (1949).
2. P. Shewmon, *Acta Met.* **8,** 606 (1960).
3. C. P. Flynn and A. M. Stoneham, *Phys. Rev.* **B1,** 3966 (1970).
4. H. K. Birnbaum and C. Baker, *Ber. Bun. Ges.* **76,** 827 (1972).
5. G. H. Vineyard, *J. Chem. Phys.* **3,** 121 (1957).
6. H. K. Birnbaum and C. A. Wert, *Ber. Bun. Ges.* **76,** 806 (1972).

225

CHAPTER 3

1. R. O. Simmons and R. W. Balluffi, *Phys. Rev.* **117,** 52 (1960).
2. P. R. Beaman, R. W. Balluffi, and R. O. Simmons, *Phys. Rev.* **A134,** 432 (1964).
3. J. Bauerle and J. Koehler, *Phys. Rev.* **107,** 1493 (1957).
4. J. Hudson and R. Hoffman, *Trans. AIME* **221,** 761 (1961).
5. R. E. Howard, *Phys. Rev.* **144,** 650 (1965).
6. N. L. Peterson, *Phys. Rev.* **136,** A568 (1964).
,7. J. G. Mullen, *Phys. Rev.* **121,** 1649 (1961).
8. J. R. Manning, *Phys. Rev.* **128,** 2169 (1962); **136,** A1758 (1969).

CHAPTER 4

1. J. R. Manning *Phys. Rev.* **139,** A126 (1965); **139,** A2027 (1965).
2. H. M. Gilder and D. Lazarus, *Phys. Rev.* **145,** 507 (1966).
3. J. R. Manning, *Acta Met.* **15,** 817 (1967).
4. R. E. Howard and J. R. Manning *J. Chem. Phys.* **36,** 910 (1962).
5. P. S. Ho and H. B. Huntington, *J. Phys. Chem. Solids,* **27,** 1319 (1966).
6. A. R. Grove, *J. Phys. Chem. Solids* **20,** 88 (1961).
7. P. S. Ho, T. Helenkamp, and H. B. Huntington, *J. Phys. Chem. Solids,* **26,** 251 (1965)

CHAPTERS 5 AND 6
References included.

CHAPTER 7

1. M. Wüttig and H. K. Birnbaum, *Phys. Rev.* **147,** 495 (1966).
2. R. F. Canon and J. P. Stark, *J. Appl. Phys.* **40,** 4366 (1969).
3. R. E. Hoffman and D. Turnbull, *Acta Met.* **2,** 419 (1954).
4. J. C. Fisher, *J. Appl. Phys.* **22,** 74 (1962).
5. T. R. Anthony, R. Dyson, and D. Turnbull, *J. Appl. Phys.* **37,** 2925, 2370 (1966).

APPENDIX E

PROBLEMS

CHAPTER 1

1. Consider fluxes of matter only in the x direction entering and leaving a specimen volume element of differential length Δx and derive the conservation equation

$$\frac{\partial C_i}{\partial t} = -\frac{\partial J_i}{\partial x}$$

where C_i is concentration and J_i is the flux of component i.

2. For diffusion in one dimension with a variable diffusion coefficient $D(x)$, determine the steady-state solution to Fick's second law,

$$\frac{\partial C}{\partial t} = \frac{\partial}{\partial x} D(x) \frac{\partial C}{\partial x}$$

for a specimen of length L, assuming the terminal concentrations differ.

3. For cylindrical coordinates with constant diffusivity show that the steady-state form for Fick's second law is

$$\frac{\partial^2 C}{\partial r^2} + \frac{1}{r} \frac{\partial C}{\partial r} = 0$$

Find the general solution to this equation.

4. For a semi-infinite specimen where the concentration satisfies

$$\frac{\partial C}{\partial t} = D \frac{\partial^2 C}{\partial x^2}$$

227

determine the solution assuming $C(x, t = 0) = 0$ and

$$\int_0^t J(x = 0) \, dt = Q_0(1 - e^{-\lambda t})$$

one form for the answer is

$$C(x, t) = Q_0\lambda \int_0^t \frac{\exp-[\lambda(t-\Gamma)+x^2/4D\Gamma]}{\sqrt{(\pi D\Gamma)}} \, d\Gamma$$

5.   With chemical reactions, one can get coupled differential equations as follows:

$$\frac{\partial C_0}{\partial t} = D_0\frac{\partial^2 C_0}{\partial x^2} - \lambda C_0$$

$$\frac{\partial C_i}{\partial t} = D_i\frac{\partial^2 C_i}{\partial x^2} + \lambda C_0$$

Assuming $C_0(x, t = 0) = Q_0\delta(x)$, show that the solution to the first of these equations is

$$C_0 = \frac{Q_0 \exp[-(\lambda t + x^2/4D_0 t)]}{\sqrt{(\pi D_0 t)}}$$

6.   With regard to problem 5, assuming $C_i(x, t = 0) = 0$ and $D_i \gg D_0$ so that $D_0/D_i \to 0$, determine the solution to the second equation using the solution to the first equation as given. One form for the solution is

$$C_i = Q_0\lambda \int_0^t \frac{\exp-[\lambda(t-\Gamma)+x^2/4D_i\Gamma]}{\sqrt{(\pi D_i\Gamma)}} \, d\Gamma$$

7.   Consider diffusion on a surface whose atoms are arranged in a square array. Reduce the diffusivity tensor by showing that $D_{12} = D_{21} = 0$, $D_{11} = D_{22}$.

8.   Let $D_k = 10^{-9}$ cm^2/sec and $\alpha = 1$ cm^{-1} and numerically integrate Eq. 1.27 to find $C_K(x)$ for an anneal of $t = 10$ hr, for $t = 1000$ hr. Discuss the means to determine D_K^* and α from experimental data based on your numerical results.

9.   From Eq. 1.34, show that C_K is a maximum or $dC_k/dx = 0$ at $x = \alpha D_K^* t$. Discuss this as a method to determine α. What other data would be needed to determine both α and D_K^*?

10.   Let $f(x) = C_1$ be a constant for $0 < x < \ell/2$, and $f(x) = C_2$ be another constant for $\ell/2 < x < \ell$, where $C_1 > C_2$ and determine the Fourier coefficients in Eq. 1.59. Hint: Let $\psi = (C_k - C_2)/(C_1 - C_2)$ and solve the problem for $\psi(x, t)$; $0 \leq \psi \leq 1$.

11. The following serial sectioning data were obtained for the diffusion of a thin layer of radioactive Ag vapor deposited on an aluminum cylinder and annealed for 25 hr.

Count rate per section	Section thickness (microns)
5010	101
3986	112
2498	99
1396	96
563	110

Determine the concentration (relative units) as a function of penetration distance x. Least-square fit the resultant data and determine D_K^* from the slope of $\ln c$ versus x^2.

CHAPTER 2

1. From Eqs. 2.3 and 2.4a, show the validity of Eq. 2.4b.

2. From Eq. 2.6, derive Eqs. 2.7a and b.

3. Consider the statistical mechanics of a dilute binary alloy with an fcc lattice and an interstitial solute. Neglect the possibility of pairs forming and determine the chemical potential of the solute. Compare your answer with Eq. 2.55 for the case where $C_0 \to 0$ and $I_0 \to 0$.

4. Consider the statistical mechanics of a dilute binary fcc substitutional alloy and determine the chemical potential of the solute and the solvent. Assume the solute is so dilute that no solute pairs form; neglect the presence of vacancies.

5. Discuss the similarities and differences between Eqs. 2.31 and 2.34. Hint: Distinguish between long- and short-range forces.

6. The Soret gradient is a long-range steady-state concentration gradient established as a result of a temperature gradient in an alloy. Assume a homogeneous substitutional solute distribution and derive the Soret gradient of the interstitial solute using Eq. 2.48a for the absence of an electric field.

7. Consider the substitutional solute and solvent to be immobile and derive the Soret gradient (see problem 6) from Eq. 2.93. From the two flux equations, 2.48 and 2.93, determine the phenomenological coefficients using the chemical potential as given in Eq. 2.55.

8.   Extend the phenomenological equation, 2.93, to the case in which an electric field can move matter. Check your answer using Eqs. 2.48a and 2.47 for the case of an interstitial solute by determining the phenomenological coefficients.

9.   Consider the potential, ϕ_i, and the frequency, ν_i, in Eq. 2.6 to be dependent on atomic volume. Discuss an equation of state for the crystal using the thermodynamic relation $P = \partial A/\partial V|_{N,T}$ for a single component crystal.

10.   Assume that an infinitely dilute binary interstitial solute in an fcc solvent resides in tetrahedral sites. Assume there are no substitional solutes present, no vacancies, and the solute concentration is sufficiently low that no interstitial pairs form. Derive the flux equation using Eq. 2.34 for the case of possible electric field, temperature gradient, and concentration gradient. Octahedral interstitial sites were assumed in the chapter development of Eq. 2.48a; your equation should be similar to this, except that $A = B = 0$.

CHAPTER 3

1.   Derive Eq. 3.87 for $\nu_{\pi-\alpha}$ using Eqs. 3.79, 3.85, 3.89, and 3.90.

2.   Consider the statistical mechanics of a dilute substitutional binary alloy in an fcc lattice. Ignore solute pairs but permit an interaction between vacancies and solute atoms of binding energy $E_b/K_BT = \ln w_4/w_3$. Derive Eq. 3.17, where z is 12. You will need to assume that the concentration of vacancies as very small.

3.   For a bcc structure with nearest neighbor jumps, consider the three classes of vacancy sites for diffusion in the $\langle 100 \rangle$ direction, determine the matrix elements for the matrices A_1, A_2, and A_3. Calculate P_{11} and P_a. Determine the correlation factor f using Eq. 3.45.

4.   It is possible to directly calculate $P_a - P_{11}$ by Eq. 3.58 for the case in which the vacancy is random if it reaches the plane containing the tracer. When the vacancy reaches the plane containing the tracer, it has an equal likelihood of contributing to P_a and P_{11}. Thus one can calculate $P_a - P_{11}$ directly with Eq. 3.58, with A equal to a 1×1 matrix. With this 1×1 approach to diffusion in fcc along the $\langle 100 \rangle$ direction, calculate $P_a - P_{11} = w_2/(w_2 + 2w_1 + 7w_3)$, and then show that f is that given in Eq. 3.62. See Chapter 5 for further discussion of this method.

5.   Using the approach in problem 4, calculate f for a bcc structure and compare your result with the answer for problem 3.

6. From problem 3 and Eq. 3.79, determine the average jump frequency and diffusivity for $\langle 100 \rangle$ diffusion in a bcc structure using the random walk approach; that is, $D = 4b^2 w_2(w_4/w_3)f$.

7. Using random walk of a vacancy, derive the steady-state concentration of vacancies next to a solute atom in bcc for the nearest-neighbor interaction model. Use the approach of Eqs. 3.81–3.83.

8. For isotropic diffusion in a crystal, that is, in a cubic structure

$$D_K^* = \frac{1}{6} \lim \frac{\langle R_n^2 \rangle}{t_n}$$

where R_n is the radial distance from the origin for n jumps of a tracer in any direction. Letting $\Gamma = \lim_{n \to \infty} n/t_n$, and using the above, derive the equation

$$D_K^* = \frac{\lambda^2}{6} \Gamma f$$

where f is the correlation factor, and λ is the absolute magnitude for the jump distance. Use a crystal structure where only one λ exists. Show that the correlation factor may be expressed as

$$f = \lim_{n \to \infty} \left\{ 1 + \frac{2}{n} \sum_{j=1}^{n-1} \sum_{i=1}^{n-j} \frac{\langle R_i R_{i+j} \rangle}{\lambda^2} \right\}$$

9. Determine the correlation factor for surface diffusion along a basal plane of a hexagonal close packed crystal.

10.† Using a high speed computer, second- and then third-nearest neighbors to the tracer as configurations in A, and an fcc structure, validate Eqs. 3.63 and 3.64.

11.† Extend Eq. 3.83 to a second-nearest neighbor calculation with the jump frequencies W_i, $i = 0, \cdots 4$ by expanding the dimensionality of the matrix. Use numerical techniques.

12.† Define a set of jump frequencies consistent with second-neighbor and first-neighbor vacancy-solute interaction for a dilute substitutional alloy. Show the influence of second neighbor interactions on the correlation factor giving violations to Eqs. 3.63 and 3.64.

13.† With the second neighbor model of problem 12, show the influence of second-neighbor interactions on either Eqs. 3.16 or 3.83 by giving corrections accountable to the second-neighbor attractions.

† Problems marked with an asterisk are considerably more involved but the physical results are thought to be instructive.

CHAPTER 4

1. Show that Eq. 4.55 follows from Eq. 4.54 when $A = A_0 + \delta A$ and nonlinear terms are omitted. The A may not commute with δA.

2. Assume vacancy equilibrium on class 3 sites for an fcc alloy and derive the expression for $KT\bar{\mu}/D$ using classes 1 and 2 sites only. Use the electric field as a driving force so that your answer compares with Eq. 4.84.

3. The correlation factor for a substitutional solute with nearest-neighbor sites interacting only in bcc is $f = 7w_3/(2w_2 + 7w_3)$. Calculate $KT\bar{\mu}/D$ for an electric field in a bcc structure.

4. Determine by random walk the matrix S_{+0} for a bcc structure and subsequently that $f = 1 - 2S_{+0}$; see Eq. 4.66.

5. Expand Eq. 4.72 into the form $A = A_0 + \delta A$, where δA is linear in the field strength.

6.† Consider an extended number of neighboring vacancy sites to the solute in fcc (i.e., second- and third-neighbor) and show the validity of the equation**

$$\frac{KT\mu}{D} = q_i \frac{2w_1(1 - 2q_s/q_i) + w_3\{7F + 6q_s/q_i - (w_4 - w_0)/w_4 14(1 - F)q_s/q_i\}}{2w_1 + 7Fw_3}$$

numerically, where F is given by Eq. 3.64. This calculation may be simplified using the procedure indicated in problem 2 by assuming that the vacancies or the plane of the tracer are at equilibrium and ignorable.

7.† Consider the second-neighbor model introduced in problems 3.12 and 3.13. Determine the influence of second neighbor vacancy bonding on $KT\bar{\mu}/D$ numerically for an electric field in one fcc lattice. Compare your results with the equation given in problem 6.

CHAPTER 5

1. Consider the two distinguishable divacancy configurations for diffusion in the $\langle 100 \rangle$ direction in fcc. Let the vacancies be infinitely bound to the tracer. Show that the partial correlation factors are $\pm\frac{1}{3}$. (See Eq. 5.29.)

2.† Assuming a mechanism for solute diffusion by infinitely bound vacancy pairs (divacancies), calculate $KT\bar{\mu}/D$ for an electric field using a

---

** J. R. Manning, *Diffusion Kinetics for Atoms in Crystals*, Von Nostrand, 1968.
† Problems marked with an asterisk are considerably more involved but the physical results are thought to be instructive.

jump frequency model similar to that for single vacancy-solute interactions (W_i, $i=0, \cdots 4$). As a check on your calculations, $KT\bar{\mu}/D = q/f$ when one has solvent tracer diffusion in the field. A reasonable number of configurations would be those in which at least one vacancy neighbors the tracer; in this instance $f \cong 0.51$.

CHAPTER 6

1.   Using Eqs. 6.27 and 6.28, calculate the correlation factor for diffusion in an fcc lattice along the $\langle 111 \rangle$ direction with a single vacancy jump mechanism (W_i, $i=0, \cdots 4$). Compare your results with the correlation factor for diffusion in the $\langle 100 \rangle$. Do both calculations with those configurations in which the vacancy is a nearest neighbor to the solute. (For the $\langle 100 \rangle$ diffusion direction, f is given in Chapter 3.)

2.† For the $\langle 111 \rangle$ diffusion direction, calculate $KT\bar{\mu}/D$ with an electric field for an fcc lattice with diffusion by a single vacancy mechanism. Compare your results with those of Chapter 4.

3.† Calculate $KT\bar{\mu}/D$ for diffusion with an electric field by a single vacancy mechanism for tracer motion parallel and perpendicular to the basal plane in the hcp lattice. Compare the results.

CHAPTER 7

1.   Discuss the reasons why one may neglect the presence of dislocations for the diffusion of interstitial solutes.

2.   Derive Eq. 7.10.

3.† Extend the A_4 matrix for the diffusion in a dislocation to third-neighbor sites and confirm the corrections introduced by F in Eqs. 7.27 and 7.28. Hint: Eq. 7.22 is valid if A_4 is $M \times M$, R is $N \times M$ and Q is $M \times N$, where N is the number of tracer jump configurations and $M \geqslant N$.

4.   Discuss alternate interpretations for the near surface effect.

---

† Problems marked with an asterisk are considerably more involved but the physical results are thought to be instructive.

INDEX